A Hunter-Gatherer's Guide to the 21ˢᵗ Century

A Hunter-Gatherer's Guide to the 21st Century

Evolution and the Challenges of Modern Life

HEATHER HEYING
and **BRET WEINSTEIN**

Portfolio / Penguin

PORTFOLIO / PENGUIN
An imprint of Penguin Random House LLC
penguinrandomhouse.com

LIBRARY OF CONGRESS CATALOGING-IN-PUBLICATION DATA
Names: Heying, Heather E. (Heather Elizabeth) author. | Weinstein, Bret, author.
Title: A hunter-gatherer's guide to the 21st century : evolution and the challenges
    of modern life / Heather Heying and Bret Weinstein.
Description: First edition. | New York : Portfolio, 2021. | Includes bibliographical
    references and index.
Identifiers: LCCN 2021005457 (print) | LCCN 2021005458 (ebook) |
    ISBN 9780593086889 (hardcover) | ISBN 9780593086896 (ebook)
Subjects: LCSH: Human ecology. | Human evolution—Social aspects. | Civilization,
    Modern—Social aspects. | Hunting and gathering societies. | Biodiversity.
Classification: LCC GF50 .H48 2021 (print) | LCC GF50 (ebook) | DDC 304.2—dc23
LC record available at https://lccn.loc.gov/2021005457
LC ebook record available at https://lccn.loc.gov/2021005458

Printed in the United States of America
4th Printing

Book design and map illustration by Daniel Lagin

To Douglas W. Heying and Harry Rubin,
who saw so much, so early, and with such clarity

Contents

Introduction

IN 1994, WE SPENT OUR FIRST SUMMER IN GRADUATE SCHOOL AT A TINY
field station in the Sarapiquí region of Costa Rica. Heather was studying
dart-poison frogs; Bret homed in on tent-making bats. Every morning we
did fieldwork in the rain forest, where it was green and lush and dark.

We remember a particular afternoon in July. A pair of macaws flew
overhead, silhouetted against the sky. The river was cool and clear, and
trees full of orchids crowded the bank. It was a perfect antidote to the
sweat and heat of the day. On beautiful afternoons like this one, we would
walk across the paved road that went all the way to the capital, onto a
smaller dirt road, and cross a steel bridge that spanned the Río Sarapiquí,
to take a swim at the beach below.

We paused on the bridge to admire the view: the river wending its way
between walls of forest, a toucan flying between trees, the distant calls of
howler monkeys. A local man whom we did not know approached and
began talking to us.

"You are going to swim?" he asked, pointing at the sandy bank where
we were headed.

"Yes."

"Today there was rain in the mountains," he said, pointing to the south.
The river's source was in those mountains, in the cordillera. We nodded.
Earlier, we had seen the thunderclouds above the mountains from the field
station. "Today there was rain in the mountains," he said again.

"But no rain here," one of us said, laughing lightly, not knowing how to make small talk in a language we weren't fluent in, while standing on a bridge, eager to swim.

"Today there was rain in the mountains," he said a third time, more emphatically. We looked at each other. Perhaps it was time to take our leave, to walk down to the river and get in the water. The sun was now directly on us. It was desperately hot.

"Okay, see you later," we said, waving, moving on. We were barely fifty feet from getting in the water.

"But the river," the man said to us, now with some urgency.

"Yes?" we asked him, confused.

"Look at the river," he said, pointing. We looked down. It looked like the river always did. Running fast and clean, smooth and . . .

"Wait," said Bret. "Is that a whirlpool? That wasn't there before." We looked at the man again, questions in our eyes. He pointed again to the south.

"Today there was a *lot* of rain in the mountains." He moved his focus back to the river. "Look at the water now."

In the moments we had been looking away, the water had come up visibly. It was moving chaotically, roiling. It had changed color, too—from dark and calm, it had become pale and filled with silt. In short order, it was filled with more than that.

The three of us stood transfixed, as the river rose spectacularly, many feet in just a few minutes. The beach disappeared under a huge volume of rushing water. Anyone on it would have been swept away. Debris, including several logs, began to hurtle past. Anything that hit that new whirlpool disappeared, then shot back up beyond the bridge.

The man turned around and began to walk off the way that he had come. He was a campesino, a farmer, but we didn't know where he was from, or how he knew that we were there, about to descend to what could easily have been our deaths.

"Wait," Bret called, then realized that we had nothing to offer him but gratitude. We literally had nothing on us but our clothes. "Thank you," we said. "Thank you so much." And Bret took off his shirt and gave it to the man.

"Really?" the man asked, as Bret held out his shirt.

"Really," Bret confirmed.

"Thank you," he said, accepting the shirt. "Good luck. And remember to think about the rain in the mountains." With that, he left.

We had been living by that river for a month, swimming in it nearly every day, sometimes alongside local people. Suddenly, we felt like strangers. We'd mistaken our few experiences swimming in the river for the wisdom of actually knowing a place. How could we have been so wrong?

At no other time in history has it been possible to think that you are a local but to be so lacking the deep knowledge of a place that keeps you safe during rare events. We moderns struggle to grasp this gap in our knowledge for many reasons. For starters, we no longer rely on tight-knit communities or a deep understanding of local terrain like humans did until recently. Given how easy it is to move from place to place with relative ease, many people tend not to stay in one locale for long at all. The facts of our individualistic lifestyles and transience tend never to strike us as odd, simply because we've neither seen nor can imagine an alternative to the world we live in right now: one where abundance and choice are ubiquitous, we rely on global systems too complex to understand, and everyone feels safe.

Until they don't.

The truth is, safety too often proves to be a facade: products on supermarket shelves turn out to be dangerous; a frightening diagnosis reveals weaknesses in a health-care system too focused on symptoms and profits; an economic downturn stresses a disintegrating social safety net; legitimate concerns about injustice become excuses for violence and anarchy while civic leaders offer pablum rather than solutions.

The problems that we face today are both more complex and simpler than experts make them seem. Depending on whom you've asked, you may have heard that we are living in the best, most prosperous time in human history. You may have also heard that we are living through the worst and most dangerous time. You may not know which side to believe. What you do know is that you can't seem to keep up.

Over the past few hundred years, developments in technology, medicine, education, and so much more have accelerated the rate at which we are exposed to change in our environments—including our geographic, social, and interpersonal environments. Some of this change has been wildly positive, but hardly all, and other changes appear positive but have consequences so devastating that, once discovered, we struggle even to conceptualize them. All of this has encouraged the postindustrial, high-tech, progress-oriented culture we live in now. This culture, we propose, partially explains our collective troubles, from political unrest to wide-spread failing health and broken social systems.

The best, most all-encompassing way to describe our world is **hyper-novel**. As we will show throughout the book, humans are extraordinarily well adapted to, and equipped for, change. But the rate of change itself is so rapid now that our brains, bodies, and social systems are perpetually out of sync. For millions of years we lived among friends and extended family, but today many people don't even know their neighbors' names. Some of the most fundamental truths—like the fact of two sexes—are in-creasingly dismissed as lies. The cognitive dissonance spawned by trying to live in a society that is changing faster than we can accommodate is turning us into people who cannot fend for ourselves.

Simply put, it's killing us.

In part, this book is about generalizing this message to all aspects of our lives: when it rains in the mountains, stay out of the river.

Many people have attempted to explain the cultural dissolution we face, but most have failed to provide a holistic explanation that not only exam-ines our present, but also looks back into our past—our whole past—and into the future. We are evolutionary biologists who have done empirical work on sexual selection and the evolution of sociality, and theoretical work on the evolution of trade-offs, senescence, and morality. We are also married to each other, have a family together, and have often been side by side while exploring many parts of the globe. Well over a decade ago, when we were still college professors, we began formulating the idea for this book. We stood on the shoulders of giants—our mentors and senior col-leagues, as well as many intellectual ancestors whom we never met—but were also building curriculum that was unlike any that came before. We forged new paths, and posited new explanations for patterns, both old and new. We came to know our undergraduate students well, and as they engaged our curricula, they asked questions across domains: What should I be eating? Why is dating so difficult? How do we create a more just and free society? The common threads throughout these conversations—in classrooms and labs, in jungles and around campfires—were logic, evolution, and science.

Science is a method that oscillates between induction and deduction— we observe patterns, propose explanations, and test them to see how well they predict things we do not yet know. We thus generate models of the world that, when we do the scientific work correctly, achieve three things: they *predict more* than what came before, *assume less*, and come to *fit with one another*, merging into a seamless whole.

Ultimately, in this book and with these models, we seek a single, consistent explanation of the observable universe that has no gaps, takes nothing on faith, and rigorously describes every pattern at every scale. This goal almost certainly cannot be attained, but there is every indication that it can be approached. Though we may glimpse this end point from our modern perch, we are a long way from reaching the limits of what can be known.

That said, we are much closer to the goal in some areas than in others. In physics we seem tantalizingly near a "theory of everything,"[1] which really means a complete model of the least complex, most fundamental layer of explanation. As we move up in complexity, things become less and less predictable. Near the top of the stack we reach biology, where processes inside even the simplest living cells are nowhere near fully understood. Things only get more complex from there. As cells begin to function in coordinated ways, becoming organisms made up of distinct tissues, the degree of mystery compounds. The unpredictability jumps again in animals, governed by sophisticated neurological feedbacks that themselves investigate and predict the world, and once again as animals become social and begin to pool their understanding and divide their labor. Nowhere are we more regularly stumped than we are in understanding ourselves. We

Homo sapiens are brimming over with profound mysteries—surrounded by paradoxes born of the very things that make us distinct from the rest of the biota.

Why do we laugh, cry, or dream? Why do we mourn our dead? Why do we make up stories about people who never lived at all? Why do we sing? Fall in love? Go to war? If it's all about reproduction, why do we take so many years to get on with it? Why are we so picky about with whom we choose to do it? Why are we fascinated by the reproductive behavior of others? Why do we, sometimes, choose to impair and disrupt our own cognition? The list of human mysteries is endless.

This book will address many of those questions. It will bypass others. Our primary aim here is not to simply answer questions but to introduce you to a robust scientific framework for understanding ourselves, one we have developed over decades of study and teaching on the topic. It is not a framework you will find elsewhere; we developed it by working from first principles as much as possible.

First principles are those assumptions that cannot be deduced from any other assumption. They are foundational (like axioms, in math), and so thinking from first principles is a powerful mechanism for deducing truth, and a worthy goal if you are interested in fact over fiction.

Among the many benefits of first principles thinking is that it helps one avoid falling prey to the naturalistic fallacy,[2] which is the idea that "what is" in nature is "what ought to be." The framework that we present here is built to free us from these sorts of traps. It is intended to allow us humans to make sense enough of ourselves that we can, at a minimum, protect ourselves from self-inflicted harm. In this book, we will identify the most large-scale problems of our time, not through the limiting, divisive lens of politics, but through the indiscriminate lens of our evolution. One of our hopes is that we can help you to see through the noise of our modern world and become a better problem solver.

Modern *Homo sapiens* arose approximately two hundred thousand years ago, the product of 3.5 billion years of adaptive evolution. We are, in most ways, a generic species. Our morphology and physiology, though staggering and marvelous when considered in isolation, are not special when compared to those of our nearest relatives. But we, uniquely, have transformed the globe and become a threat to the planet on which we still thoroughly depend.

We might have called this book *A Postindustrialist's Guide to the 21st Century*. Or An Agriculturalist's Guide. Or A Monkey's Guide, or A Mammal's Guide, or A Fish's Guide. Every one of those represents a stage of evolutionary history to which we have adapted, and from which we carry evolutionary baggage: our Environment of Evolutionary Adaptedness, or EEA, to use the term of art. In this book, we speak to our Environments of Evolutionary Adaptedness—which is to say, not just the EEA of the title, such as the African grasslands and woodlands and coasts on which our ancestors were hunter-gatherers for so long, but the many other EEAs to which we are adapted. We emerged onto land as early tetrapods; became lactating, fur-bearing mammals; developed dexterity with our hands and visual acuity as monkeys; grew and harvested our own food as agriculturalists; and live cheek to jowl with millions of anonymous others as postindustrialists.

We chose to include *hunter-gatherer* in the title of the book because our recent ancestors spent millions of years adapting to that niche. This is the reason so many people romanticize this particular phase of our evolution. But there was not just one hunter-gatherer way of life, any more than there is one mammalian way of life, or a single way to farm. And we are not adapted only to being hunter-gatherers—we also adapted, long ago, to being fish; more recently, to being primates; and most recently, to being postindustrialists. All of these are part of our evolutionary history.

This wide-ranging view is necessary if we are to understand the biggest problem of our time: Our species' pace of change now outstrips our ability to adapt. We are generating new problems at a new and accelerating rate, and it is making us sick—physically, psychologically, socially, and environmentally. If we don't figure out how to grapple with the problem of accelerating novelty, humanity will perish, a victim of its success.

This is a book not only about how our species is in danger of destroying our world. It is also about the beauty humans have discovered and created, and how we can save it. An irrefutable evolutionary truth undergirding this book is that humans are excellent at responding to change and adapting to the unknown. We are explorers and innovators by design, and the same impulses that have created our troublesome modern condition are the only hope for saving it.

A Hunter-Gatherer's Guide to the 21st Century

Chapter 1

The Human Niche

It was the best of times, it was the worst of times, it was the age of wisdom, it was the age of foolishness, it was the epoch of belief, it was the epoch of incredulity, it was the season of Light, it was the season of Darkness, it was the spring of hope, it was the winter of despair, we had everything before us, we had nothing before us.

Charles Dickens, from the opening lines of *A Tale of Two Cities*, published in 1859, the same year that Charles Darwin published *On the Origin of Species*

BERINGIA WAS A LAND OF OPPORTUNITY, A VAST AND OPEN GRASSLAND. A landmass four times the size of California that connected Alaska to the east with Russia to the west, Beringia was not merely a temporary land bridge, a passage between Asia and the Americas. People did not scurry across, the rising waters lapping at their feet, nor was it a lifeless plain. Life was surely difficult, but for thousands of years, Beringia supported a population of people who made their home there.[1]

The people who came to Beringia were fully modern in every genetic and physical sense. They came from the west, from Asia, and for a long time there was a barrier of ice at Beringia's eastern edge. So they settled there, and many generations passed. As the world warmed, though, the ice began to melt, sea levels rose, and Beringia began to disappear, the coastline encroaching on what had been home. Where to go?

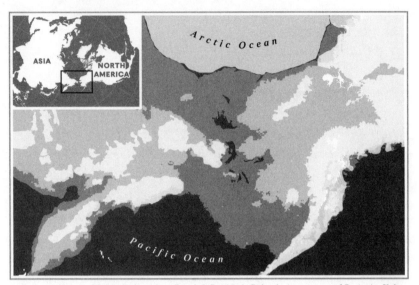

Artist's rendering of Beringia based on Bond, J. D., 2019. *Paleodrainage map of Beringia.* Yukon Geological Survey, Open File 2019-2.

Some Beringians no doubt went west, back to Asia, from where all of their ancestors had come, a land that may have lived in myth and collective memory. Perhaps in the intervening years, newer arrivals had come from there, too, and brought with them updated stories of what their home to the west was like.

As sea levels rose on the Beringians, some headed east, into a land that no humans had ever seen before. These were the first Americans. Probably the Beringians traversed that northern part of the west coast by boat.[2] The ice was still there, but there would likely have been ice-free refugia peppering the coast, places where local animals concentrated, places that may have acted as stepping-stones for those first Americans.[3]

This was, best estimates now suggest, at least fifteen thousand years ago,[4] and possibly far deeper into history even than that. Depending on what that ice sheet looked like, perhaps they couldn't make permanent landfall until they got as far south as what is now the city of Olympia, in Washington State. It was there that the glaciers ended. South of Olympia, and east, landmasses unimaginable in their scope and variety, full of verdant and beautiful landscapes and delicious and charismatic organisms, but no people, were about to be explored by humans for the first time.

It was a risky move. The whole thing was incredibly risky. None of the choices seemed good. Go back to the west, to a land already occupied by people who undoubtedly have opinions about newcomers? Head east, to a land nobody knows anything about? Or stay in place as Beringia disappears into the sea? Nobody who survived chose the third option. Go back to what your people once knew, a place vetted and abandoned by your ancestors, a place known to be full of competitors . . . or explore someplace completely new? Both are legitimate choices, both have distinct risks, distinct advantages and disadvantages. These are, likewise, the options in our modern world.

The descendants of the Beringians would come to populate the Americas in total isolation from all human populations in the Old World. They arrived before any humans on Earth had invented written language or agriculture; independent of any input from their Old World relatives, they innovated these things from scratch. Their lineage would discover hundreds of new ways of being human, and rise to an estimated population of fifty million to one hundred million before Spanish conquistadors brought the Old and New World populations into violent reconnection many thousands of years later.

We do not know for certain what the journey to the New World looked like. Perhaps the first Americans were even earlier, not making permanent homes in Beringia at all, instead circumnavigating the Pacific, clockwise, in boats.[5] What we do know is that the New World presented challenges that the first Americans had never seen before. And this story of Beringia, even if true only at the metaphorical level, is instructive of what it is to be human. It is an apt if incomplete metaphor for the situation that humanity is in today. We too find ourselves in a failing land. We too must seek new opportunities to save ourselves. And we too do not yet know what exploration will yield.

Early Americans found themselves in an immense landscape of unknown hazards and opportunities. With ancestral knowledge that was ever less relevant as a guide, the challenges of navigating this new world would have been immense. And yet they succeeded spectacularly. The question that we ask, and which is most pertinent to our modern situation, is *How?* The answer will be found, in large measure, in understanding what it is to be human.

Several generations in, sitting around the fire at night, a bit hungry because berry season was past its peak and the deer had grown scarce, one of these early Americans, call him Bem, might have observed that bears seem to sustain themselves on the fish, so why can't we?[6] Bem didn't know much about fish, though, not like Soo, who had spent many days at river's edge, watching fish, and had insight about how the fish behaved. Soo's insight about fish was, heretofore, not one she had shared, nor one that had seemed to her likely to have any value to her people. Soo, in turn, may not have had latent engineering skills, as Gol did, and Gol may have lacked Lok's talent in experimenting with making rope. When so many people with distinct talents and insights come together around a camp fire to discuss a shared problem, the spark of innovation can spread quickly.

Most of the best ideas that our species has generated, the most important and powerful ideas, have been the result of a group of people who had different but consilient talents and vision, non-overlapping blind spots, and a political structure that allowed for novelty. Gathered around the fire at the threshold of two continents new to humanity, many insightful observers and engineers, crafters of tools and synthesizers of information, came together and learned, or relearned, how to fish salmon from rivers, what bulbs were safe to eat and how to identify them, and how to transform trees into shelter. Those populations had keepers of the flame, too: holders of tradition, individuals who would tell the story later, perhaps when a move was necessitated by the failure of a local salmon run and all of the original innovators were gone.

What all were Bem or Soo or Gol or Lok doing, exactly? They were innovating, as part of and on behalf of their people. They were testing hypotheses, creating narrative, building material and culinary traditions. They were being human.

The Human Paradox

Twenty-first-century people face opportunities and dilemmas similar to those of the original people of the New World. Innovations in technology and science have allowed us to enter new, previously unimagined realms. But unlike the Beringians, we do not have an ancestral land to even consider returning to, because our actions affect the entire planet. We have hunted and gathered, cultivated and machined our way around the globe, transforming the earth in our wake, bending landscapes to our will and pushing many to the brink of collapse.

Some look back at our species' successes, such as the success of the Beringians, and imagine that we can master nature, that we are in control of it. But we are not, and we never will be.[7] The consequences of this bad assumption account for many of our problems today. The only way to course correct is to understand the true nature of what we are, what we might be, and how we might apply this wisdom to our benefit.

Our species is brainy and bipedal, social and talkative. We make tools, cultivate land, produce myth and magic. We have reinvented ourselves over time and across space, over and over again, learning to dominate one habitat after another. Species are defined by many things—their form and function, their genes and development, their relationship to other species. Perhaps most important, though, species are defined by their niche: the particular way they interact with and find a way to make a living in their environment.

Given our breadth of experience and geography, what exactly is the human niche?

As our species has evolved, we seem to have escaped a fundamental law of nature: the jack-of-all-trades is the master of none. To be dominant in any one niche, a species must typically specialize, sacrificing breadth and generality. It is this need to specialize that hobbles the jack-of-all-trades, a principle so universal that it has been invoked for more than four centuries in print (one of the earliest instances being a 1592 jab at actor-turned-playwright William Shakespeare).[8] "The jack-of-all-trades is the master of none" is applied broadly, from engineering to sports to ecological science. Species are, in this way at least, like tools: the more jobs they do, the more crudely they do them.

Yet somehow here we are, jacks of nearly every trade imaginable, and simultaneously the masters of nearly every habitat on Earth. Our niche is nearly unbounded, and when we do find boundaries, we nearly immediately begin to test them. It's as if we don't believe there will ever be a final frontier.

Homo sapiens is not merely exceptional. We are exceptionally exceptional.[9] Unrivaled in our adaptability, ingenuity, and exploitative capacity, we have come to specialize in everything over the course of hundreds of thousands of years. We enjoy the competitive advantage of being specialists, without paying the usual costs of a lack of breadth.

This is the paradox of the human niche.[10]

A paradox in science is like an X on a treasure map: It tells us where to dig. Our unrivaled breadth of specialization is a paradox that marks the location of a spectacular trove, not so much of riches, but of tools. By unraveling the human paradox, we can unlock a conceptual framework that allows us to understand ourselves, and to navigate our lives with intention and skill. This book unpacks the human paradox, and describes the tools we discover there; it is also an exercise in their application.

Campfire

In our discussion of the first Americans, we have already seen one tool in this treasure trove, though it might not seem to be a tool at all. It is a campfire.

Humans have been using fire for eons. We have used it to make light and create warmth, to increase the nutritional value of food, and to keep predators at bay. We have used fire to hollow logs to make canoes, to transform landscapes to new purposes, to soften and harden metal. We have also used fire for something even more important: The campfire is a forge for ideas. A place to discuss berries, rivers, and fish. A place to share our experiences, to talk, to laugh, to cry, to deliberate over our challenges and share our successes. From this forge emerge the kinds of ideas that render humans a true superspecies, one that surfs the rules of the universe, kicking up paradoxes in its wake.

The exchange of ideas that has occurred around the hearth for millennia is more than simple communication. It is the convergence point of individuals with different experience, talent, and insight. The linking of minds is at the root of humanity's success. It doesn't matter how smart an individual is, and it doesn't matter how much they know. In nearly every case, when minds come together, the whole is greater than the sum of its parts. For the problems that humanity faces—from which bulbs are safe to eat and how to catch rabbits, to how to equalize opportunity while creating a world that is safe from existential threats—we need more than individuals processing in isolation. If we are to survive the future, we need multitudes of people plugging in and parallel processing. Joining minds in this way exponentially increases the ability of humans to solve problems.

Just as humanity broke down boundaries between niches that no other organism has broken down, so too have we broken down boundaries between individuals that nothing else has broken down so thoroughly. With regard to niches, we are a generalist species that contains individuals who are often specialists. A single ancient American may have been terrific at wayfinding, but terrible at keeping the flame. A single modern human may be terrific at rock climbing but terrible at organizing their files, or excellent with numbers but unskilled in the baking of bread. As a species, though, we are supremely good at all of these things. It is the connections between us that allow us to transcend our individual limitations, often focusing on our trade while being sustained by the specialized labor of others.

At the boundaries between individuals, we consciously innovate and share ideas, and then reify the best and most relevant of those ideas for the current moment in the form of culture. For millennia, this magic has occurred around the common campfire.

Consciousness and culture—themes that we return to in depth in the penultimate chapter of this book—are in tension with each other, and humans need both.

Conscious thoughts are those that can be communicated to others. We define *consciousness*, therefore, as "that fraction of cognition that is packaged for exchange." This is no trick. We have not chosen a definition to make an intractable question simple. We have chosen the definition at the epi-center of what people mean when describing a thought as "conscious."

One truth that emerges from understanding consciousness in this way is that it makes little sense to assume that individual consciousness evolved first, or that it is the most fundamental form of consciousness. Rather, our individual consciousness likely evolved in parallel with collective consciousness, and would become fully realized only later in our evolution. Understanding what is in the mind of another—known as theory of mind—is staggeringly useful. We see the rudiments of this capacity in many other species, and we see it extensively elaborated in a highly cooperative few, such as elephants, toothed whales (such as dolphins), crows, and many nonhuman primates. We humans are by far the most aware of one another's thoughts of any species that has ever existed, because we alone can, if we so choose, hand over the cognitive goods explicitly and with spectacular precision. We can accurately pass a complex abstraction from one mind to another by simply vibrating the air between us. It is everyday magic that usually passes without our notice.

For theory of mind to function, one needs to run an emulation of the other person within one's own head. For *me* to benefit from a comparison between what *I* think on the one hand and what I understand *you* to think on the other, I am all but required to have subjective experience of both *you* and *me*—to bring the two into a single currency. Shared consciousness is an emergent, intangible space between people, where concepts are lodged and co-cultivated. Each participant has a distinct perspective on the space, much as each witness to a physical event will have a somewhat different vantage point, but the space is a property of the collective.

Imagine two populations composed of equally smart individuals. In the first population, individuals cannot just propose ideas, but also must respond to and modify the ideas of others, and then strategize and plan how to act on them, with each individual contributing in his or her own area of specialty. The second is made up of individuals who, while full of good ideas themselves, have no ability to conceptualize what others are thinking. When these two populations are in competition with each other, there is simply no contest.

Even a rudimentary collective consciousness—what might be shared between wolves in a pack, for example, as they are hunting cooperatively—provides a staggering advantage. In lions, too, the pride is far greater than the sum of its individuals. Collective consciousness, an evolutionary innovation unlike any other, creates cognitive emergence.

Culture versus Consciousness

Consciousness is valuable for problem-solving, but it isn't so good for execution. The gymnast, the virtuoso, and the warrior all succeed by taking what they have discovered consciously and learning to apply it without explicit deliberation.[11] Transformative insights and ideas move out of the conscious layer and into the parts of us that know how to get things done. When one is *in the zone*, the conscious mind is present, but as a spectator who steers clear so as not to disrupt the flow. Behaviors become habitual and intuitive. In an individual, we might call this skill or craft. In a family or a tribe, such habits become traditions, passed efficiently from one generation to the next. Scale this up further, and we have culture.

Homo sapiens therefore oscillate between two dominant modes. When we face problems for which our prior understanding is inadequate, we become conscious. *How do we feed ourselves in this new land?* We plug our minds into a shared problem-solving space and share what we know. Then we parallel process—proposing hypotheses, providing observations, offering challenges—until we arrive at a new answer, one that an individual would rarely reach alone. If the result works well when tested in the world, it gets refined and then driven into a more automatic, less deliberative layer. This is culture. The application of culture to the circumstances for which it is adapted is the population-level equivalent of an individual being in the zone.

This model implies a few important things. When times are good, people should be reluctant to challenge ancestral wisdom—their culture. In other words, they should be comparatively conservative. When things aren't going well, people should be prone to endure the risks that come with change. They should be comparatively progressive—liberal, if you will.

This of course has a lot to say about the modern world, because for various reasons, there is little agreement at present on how well things are going. Moments before the Titanic hit the iceberg, the ship was a marvelous testament to human achievement. Moments after, it was a monument to the hazard of hubris. Too often, it is only in retrospect that the rearranging of deck chairs appears absurd. More often than not, there is no iceberg, no clear demarcation of before and after, of the moment when consciousness should become more salient than culture.

Humans break

1. niche boundaries by being both generalists and specialists.
2. interpersonal boundaries by oscillating between culture and consciousness.

The financial collapse of 2008, Deepwater Horizon oil spill, and the Fukushima Daiichi nuclear disaster are all symptoms of a civilization-level disorder, one that has no name. Let's call it the **Sucker's Folly**: the tendency of concentrated short-term benefit not only to obscure risk and long-term cost, but also to drive acceptance even when the net analysis is negative.[12] These events are evidence that we are resting on our cultural laurels and speeding toward disaster, lulled into a false sense of security—and away from collective consciousness—by the opulence of our surroundings. The sooner we recognize this, the greater the chance to divert the ship to a safe course, a puzzle we will return to in the last chapter of this book.

The answer to our earlier question, then—*What is the human niche?*—is this: Humans don't have a niche, not in the standard sense of that term. We have escaped the paradigm by mastering a different game. We have discovered how to swap out our software and replace it as the need arises by oscillating between culture and consciousness. *The human niche is niche switching.*

Humanity is the master of every trade. If we were machines, we would be ones that are compatible with many software packages. The Inuit hunter knows the Arctic, but has few of the skills needed to function in the Kalahari or the Amazon. Humans can be good at almost anything, given the proper tools and software, and human populations can be good at many things by virtue of a division of labor, but each individual person will either have to limit themself or accept the costs that come with being a generalist.

As our world becomes increasingly complex, though, the need for generalists grows. We need people who know things across domains, and who can make connections between them: not just biologists and physicists, but biophysicists; people who have switched gears and found that the tools they brought from their prior vocation serve them well in a new one. We must find ways to encourage the development of generalists. In this book, we argue that a key way to do this is to encourage a careful, nuanced understanding of what evolution is, what it has made us, and how we can resist its goals. To that end, let us first, in the remainder of this chapter, provide a few updates to evolutionary theory. The alterations that we are suggesting open a path to understanding evolution more deeply, and also to understanding ourselves, our cultures, and our species.

Adaptation and Lineage

Adaptive evolution improves the "fit" of creatures to their environment. This is well established. In a rush to make evolutionary biology an empirical science, though, biologists prioritized defining *fitness* such that it could be easily measured. We biologists settled on a definition that is almost synonymous with *reproduction*. As is the case with many assumptions that ultimately fail, the belief that *fitness* and *reproductive success* were near synonyms was wildly successful at first, enough so that generations of biologists made great headway by simply treating them as one. All else being equal, a creature that is a better fit for the environment tends to produce more offspring, and when that is the case, biologists have excellent conceptual tools for unpacking the evolutionary process that leads to it. What happens when all else isn't equal, however, and the creature with more offspring has cut corners in the pursuit of short-term fecundity? Under these conditions the ability of biologists to understand the story is compromised. If the harm done to fitness shows up quickly—if an individual animal produces many offspring, all of which perish in the winter—we will likely come to understand that it failed in an evolutionary sense. If, however, the descendants prosper for a fairly long time, but die off in the next drought, or the next Ice Age, there is a good chance biologists will botch our analysis of "success."

Fitness is indeed often about reproduction, but it is *always* about persistence. A successful population can ebb and flow through time. What a successful population can't do is go extinct. Extinction is failure. Persistence is success—and the reproduction of individuals is only one factor in the persistence equation.

But what does it mean to persist? Is it species persistence we are after? Do we count each population within the species separately? Is it an individual's descendants we should be counting? Logically, it must be all of those things, and more.

Adaptive evolution occurs as individuals compete for resources. Each individual is the beginning of a line of descent, and the period over which its descendants persist is a good proxy for its fitness. If Bem's descendants perish as the glaciers return, but Soo's descendants find their way through to the next interglacial, the latter were fitter—whether we were able to measure the difference or not.

But those two individuals were not only the starting points for lines of descent going forward. Each was also a member of many simultaneous, overlapping lines of descent stretching backward to a large collection of ancestors about whom we could say the same thing. So if fitness is about persistence, then the apropos question is, The persistence of what?

This is where we must break our sense of obligation to measure things. Adaptive evolution—the process that increases the "fit" of creatures to environment—is about all levels of descent at once. Adaptive evolution is therefore fractal, and the term that encapsulates it is *lineage*.

An individual and all of its descendants comprise a lineage. A species is a lineage descended from that species' most recent common ancestor—as too are larger clades, such as mammals, vertebrates, animals—lineages descended from those clades' most recent common ancestors.[13] Our job as evolutionary biologists is to figure out how adaptive evolution works with selection at all levels of lineage occurring simultaneously. In this book we will proceed from the premise that lineages compete, and those that are better suited to their long-term environment are favored by selection. That buys us a lot in terms of illuminating the paradoxes of human nature, but it's far from sufficient. We must also recognize that, contrary to conventional evolutionary wisdom, genes are not the only form of heritable information.

Culture evolves. Furthermore, culture evolves in tandem with the genome, and is obligated to the same objective. We need not know the degree to which, for example, sex-typical behaviors like female nesting or male bravado are transmitted culturally or genetically; the mode of transmission says nothing about the meaning of these patterns. Whether cultural, genetic, or a mixture of the two, sex roles inherited from a long line of ancestors are biological solutions to evolutionary problems. They are, in short, adaptations that function to facilitate and ensure lineage persistence into the future.

This is a hard pill for many to swallow, but the truth is that culture exists in service to the genes. Long-standing cultural traits are as adaptive as eyes, leaves, or tentacles.

In the 21st century, nearly everyone accepts that evolution has created our limbs and our livers, our hair and our hearts. Yet many people still object when evolutionary theory is invoked to explain behavior or culture.[14] Even for many scientists, this position is driven by the belief that some questions should not be asked if the answers to them might be ugly. This has led to ideologically driven censorship of ideas and research programs, which has slowed the rate at which we have enhanced our understanding of who we are, and why.

Some of what evolution has produced is, indeed, ugly: infanticide, rape, and genocide are all products of evolution. It is also true that much of what evolution has produced is beautiful: a mother's sacrifice for her child; enduring romantic love; and civilization's care for its citizens, young and old, healthy and not. A widespread lack of understanding of what it means for something to be "evolutionary" explains the concern that some people have.

Many people fear that if something is *evolutionary* it must be *immutable*. If that were true, then if something horrible is the product of evolution we would be powerless against it, and would be forced to suffer the cruelty of evolutionary fate forevermore. Fortunately, this fear is wrong. Some of what is evolutionary is nearly invariant: humans have two legs, one heart, a large brain. But the variation between individuals is also evolutionary, and strongly dependent on interactions with our environment: how long are our legs, how strong are our hearts, how interconnected are the neurons in our brains? Similarly, recognizing the evolutionary truth that women are both more agreeable than men, on average, and more anxious is neither a diagnosis of any individual nor an immutable fate. Individuals are not the same as populations.[15] We are individual members of populations, and those populations—men and women, boomers and millennials, Americans and Australians—have real psychological differences, but we are more similar than dissimilar. Those differences are the results of interactions between multiple layers of evolutionary forces. Furthermore, humans have the ability to directly plug in to one another and alter our culture, for both better and worse.

In response to the widespread confusion surrounding cultural and genetic evolution, we have developed a simple model for understanding the hierarchical nature of the forces at play. We call it the Omega principle.

The Omega Principle

Epigenetic means "above the genome." The first time either of us encountered the term was in college in the early '90s. At that time it was occasionally used by evolutionary biologists to place culture in a rigorous evolutionary context.

Culture sits "above" the genome in the sense that it shapes the way the genome is expressed. Genes describe proteins and processes that construct a body. Culture—in those creatures that have culture—has a powerful influence on where bodies go, and what they do. In this way, culture is a regulator of genome expression.

The term *epigenetic* has in more recent decades taken on a different meaning. The term is now almost exclusively used to refer to mechanisms that directly—molecularly—regulate the expression of the genome—expressing some traits while suppressing others, creating the patterns of gene expression that give the body a coherent form and function. These regulatory mechanisms, which scientists are just beginning to understand, are the key to understanding multicellular life. Without these mechanisms, all cells with a given genome would be alike, and any large collection of cells could exist only as a colony of undifferentiated cells. It is only through the tight, epigenetic regulation of gene expression that we can have an animal or a plant composed of well-coordinated, distinct, multicellular tissues.

While the meaning of the term *epigenetic* has gone through a radical transformation, from describing inherited behavior to describing only molecular switches, a strong argument can be made that the category of epigenetic phenomena actually includes both types of regulators: molecular switches are the narrow meaning of the term—epigenetic *sensu stricto* ("in the strict sense"), while the molecular switches *plus* inherited behaviors are epigenetic *sensu lato* ("in the broad sense").

Both are epigenetic, and the implication is that a single evolutionary rule governs both molecular and cultural regulators of gene expression.

Let's take a Tibetan herdsman, as an example. He has an inherited culture that constrains his behavior. His cells take different forms and do different things based on inherited patterns of gene expression. It would make no sense to imagine that the genes in his genome and the molecular regulators that adjust their expression are rivals. If the herdsman is healthy, his cells serve his evolutionary interests as a creature—the regulation of his genes evolved to enhance his fitness. His eyes, composed of many kinds of cells distributed in particular ways, see danger and opportunity. The hazards he sees are threats to his evolutionary fitness, and the opportunities constitute ways in which he might enhance it. In other words, the genes and their regulators agree on the job to be done and show no sign of tension over it. What is the job of those genes and their regulators? It is obviously evolutionary—to lodge copies of the herdsman's genes deeply into the future. No reasonable person would argue otherwise.

But many otherwise reasonable people will fail to see this relationship when it comes to the herdsman's culture. He may adhere to gender roles that stretch back thousands of years in his lineage, but it is commonly asserted in scientific circles that these cultural patterns are not likely to be evolutionary, that they are "just cultural"—as if that were a competing category.

The problem stems from the initial, 1976 presentation of memetic evolution by Richard Dawkins in *The Selfish Gene*. As Dawkins describes memes—laying the foundation for the rigorous Darwinian study of cultural adaptation—he makes a fateful error. He describes human culture as a new primeval soup,[16] in which cultural traits spread themselves much like genes do, rather than as a tool of the genome that evolved to enhance the genome's fitness.

This misunderstanding has never been properly resolved, and the nature versus nurture confusion it engenders continues to block analytical and societal progress. Asking if a particular trait is due to nature or nurture implies a false dichotomy between nature, genes, and evolution on the one side and between nurture and environment on the other. In fact, all of it is evolutionary.

The key to seeing why culture must serve genes as a fitness-enhancing tool, exactly as molecular regulators do, is found in the logic of trade-offs, a concept to which we will return throughout the book.

From the genome's perspective, culture is anything but free. In fact, nothing is more costly. The brain that picks up culture is big and energetically expensive to run; the process by which culture is transmitted is prone to error; and the content of human culture frequently blocks off fitness-enhancing opportunities—thou shalt not kill, steal, covet, lay with, etcetera. Anthropomorphizing the genome for a moment: If culture did not pay the genome back for its astronomical expense, the genome would have reason to be livid. Culture appears to waste time, energy, and resources that would otherwise be at the genome's disposal. One might get the impression that culture is effectively parasitizing the genome.

But the genome is in the driver's seat. A capacity for culture is nearly universal in birds and mammals; it has been elaborated, enhanced, and extended by genomic evolution over time; and it is at its most extreme in the world's most broadly distributed and ecologically dominant species: humans. These facts tell us that whatever culture does, it is not coming at a cost to genetic fitness. Rather, it enhances fitness in dramatic ways. If culture was not paying its way, the genes whose expression it is modifying would either go extinct or evolve to be as immune to culture as an oak tree.

In our teaching of evolution to students, we have codified our understanding of the relationship between genetic and epigenetic phenomena in what we call the Omega principle. It has two elements:[17]

Omega Principle

1. Epigenetic regulators, such as culture, are superior to genes in that they are more flexible and can adapt more rapidly.
2. Epigenetic regulators, such as culture, evolve to serve the genome.

We have chosen to use the signifier Ω (omega) to call to mind π (pi), and thus indicate the obligate nature of the relationship. Adaptive elements of culture are no more independent of genes than the diameter of a circle is independent of that circle's circumference.

From the Omega principle we derive a powerful concept: any expensive and long-lasting cultural trait (such as traditions passed down within a lineage for thousands of years) should be presumed to be adaptive.

Throughout this book, we will discuss such traits—from harvest feasts to the building of pyramids—through this evolutionary lens. We will use first principles to extrapolate what makes humans so special, and why the novelty of the modern era has made us mentally, physically, and socially unhealthy. In order to discover those principles, we must look for clues. In the next chapter, we will explore our deep history, touring the many forms we have taken, some of the many systems and abilities our ancestors innovated, and the human universals that unite us all.

Chapter 2

A Brief History of the Human Lineage

THERE ARE SEVERAL HUMAN UNIVERSALS. [1]

All humans have language. We can tell *self* from *other*, and can distinguish self as subject ("I cooked for her") from self as object ("she cooked for me"). We use facial expressions that are both general and nuanced, which include happiness, sadness, anger, fear, surprise, disgust, and contempt. We don't just use tools; we use tools to make more tools.

We live in or under shelter. We live in groups, usually with family, and adults are expected to help socialize children. Children observe elders, and copy them. We also learn by trial and error.

We have status, governed by rules stemming from kinship, age, sex, and beyond. We have rules of succession and markers of hierarchy. We engage in division of labor. Reciprocity is important, both in the positive sense—barn raising for neighbors, exchanging gifts—and in the negative—retaliation for perceived wrongs. We trade.

We predict and plan for the future, or at least we try to. We have law, and we have leaders, although both may be situational or ephemeral. We have rituals, and religious practice, and standards of sexual modesty. We admire hospitality and generosity. We have an aesthetic, which we apply to our bodies, our hair, and our environment. We know how to dance. We make music. We play.

It took so long for us to become who we are today. If you look deeply into the history of life on our planet, you can see how these universals emerged over hundreds of millions of years, and once you understand this, you will see why change, especially rapid change, isn't always such a good thing.

———

Three and a half billion years ago, give or take a few hundred million years, life blinked into existence on Earth. That organism was the common ancestor of all life on our planet, and we owe much to it, although we don't resemble it much now.

The first single-celled organism had no nucleus. It had no sex. It made its own energy, perhaps by converting sunlight into food, as modern plants do, perhaps by converting inorganic molecules, such as ammonia or carbon dioxide, into food. As we move forward in history, as our ancestors get closer to us in time, we resemble them more and more closely.

Two billion years ago, our replicating material became encased in nuclei, allowing DNA to organize itself, such that careful unpacking, at the right moments, would trigger cascading events. Much complexity hides in the timing and coding of events—and in how things are packed. The ability to pack efficiently turns out to be important, for far more than suitcases and shipping containers. We were evolving so many ways to divide labor in that era—organelles within cells separated cellular functions from one another, microtubules and motor proteins began transporting cellular material around.

Now that we had cells with nuclei, we were eukaryotes, but we still lived alone, as single cells. A long time later, we began to associate more permanently with one another, combining forces, becoming multicellular individuals, rather than clusters of cells that aggregated.[2] Specialization was crossing scales. Organelles within cells had long since innovated specialization—chloroplasts for photosynthesis, mitochondria for power—but the specialization had stopped at the boundary of the cell. Now, with multicellular organisms, life was leveling up.

Everyone who knows our deep history will have their favorite transformations, the ones that seem singularly important if anything downstream were to have a chance of happening. Perhaps you see the origin of brains, or of blood, or of bone, to be *the* evolutionary transformation on which all later innovations depended. All of them, except the earliest, hinge on conditions already created, so none of them were fated to be, in the form that we know them. In the beginning, there was the evolution of those who create their own energy. This made possible the evolution of those who take what others have created: heterotrophs, like us, who parasitize the energetic work of plants and other photosynthesizers. The particular way that we have evolved to be heterotrophs, to take the energy of others for our own—there was nothing inevitable about that.

Organisms, all of us, need to respire, to take in nutrients and excrete waste, to reproduce. The larger the organism, the more likely other things are needed, too: a plumbing system to move things around within the body; a control center—or centers—within which information is collected, interpreted, and acted upon.

More than six hundred million years ago we became multicellular individuals who steal energy from those who make it from the sun—we became animals.

Sex evolved in our lineage, and it has never gone away. Some traits blink on and off through evolutionary time—birds developed flight, and then some of them reversed course, became penguins, kiwis, ostriches.[3] Snakes lost the limbs that their—and our—ancestors had developed over tens of millions of years. Even eyes, which provide the most dominant sense in humans, switch off in some species of cave fish, which live in waters so dark that eyes are no help, all hazard. In Mexican cave fish alone, there are dozens of distinct populations of eyeless forms, living near their sighted cousins on the surface.[4]

Other traits evolve once and then stick, suggesting that their value is nearly universal. No organisms that once evolved bony internal skeletons have since evolved a lifestyle without them. The same holds for neurons and for hearts. The evolution of sex—meaning, the evolution of sexual reproduction—isn't quite so clean a story, but it is nearly so. There is one eukaryotic lineage known on earth that once had sexual reproduction and has since lost it. These are the bdelloid rotifers,[5] which are highly unusual in several ways, including their ability to survive both extreme desiccation and high doses of ionizing radiation.[6] But the lineage to which we belong is one long, uninterrupted string of sexual reproduction for at least the last five hundred million years.[7]

Early in our history as multicellular animals, some lineages branched into *sessile* forms—I will protect myself in place—others into *mobile* ones—I will rove the landscape searching for what I need, escaping from what wants me. Most of us are also bilaterally symmetrical: we have a left and a right, and the midline is a point of inflection, the views from either side a near mirror image of the other. Insects have a left and right, as do we vertebrates, although we are more closely related to sea stars than we are to insects. This reveals that even a patently useful trait like bilateral symmetry is not universally so—adult sea stars apparently gave up having a left and right in favor of radial symmetry.[8]

A Phylogenetic Tree

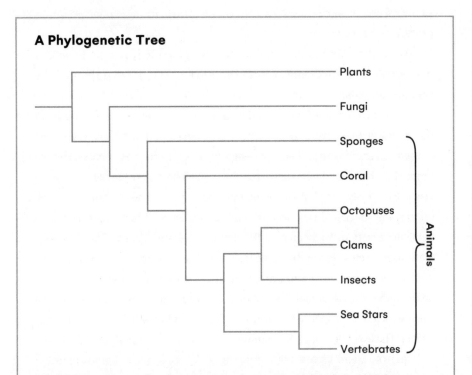

This evolutionary tree reflects our current understanding of the relationships between several extant taxa.[9] Many taxa are excluded, but the nature of evolutionary trees is that you can exclude taxa without rendering a tree untrue; it's just less complete.

This tree does *not* suggest that vertebrates are "more highly evolved" than anything else on the tree. This tree *does* suggest, among other things, that:

- Vertebrates and sea stars are more closely related to one another than either are to anything else on the tree.
- Clams and octopuses are each other's closest relatives on this tree; insects are closely related to them. Animals and fungi are more closely related to each other than either is to plants.

Five hundred million years ago we began to organize our internal activities—we evolved a single centralized heart, and brain, where before there were multiple centers for pumping and pressurizing blood, and multiple centers for neural processing. With a single brain to organize inputs, we also developed ever more ways to sense our world.

Soon enough, on a geological time scale, we had become craniates, the brainy ones, with our precious brains carefully protected within skulls. Bone hadn't evolved yet, nor jaws, so we were still limited in what we could accomplish. Yet organisms by that description persist today—lampreys, still alive, still doing fine, thank you very much—are modern representatives of those early craniates. With no jaws and no bone, their little brains work hard to find hosts to latch on to and parasitize.

Teeth and jaws evolved, and both proved useful. As did myelin, which coats the outside of neurons and allows the transmission of neurological signals to increase in speed: with myelin, our ability to move, feel, and think got faster.

By 440 million years ago, many fish were armored with sheets of bone on the outside of their bodies, but nobody yet on Earth had an internal bony skeleton. Some of the modern descendants of those fish, which have jaws and teeth but no bone, appear to be the sharks, skates, and rays.[10] Sharks, the animals of many people's nightmares, do what they do without a bone in their body. There are so many ways to be strong, to be clever, to be successful.

When bone, a molecular relative of teeth, showed up as internal skeletal material, rather than as armor, replacing the cartilage that came before it, we became Osteichthyes—bony fishes. We are also, still and forever, eukaryotes, animals, vertebrates, craniates. Group membership never disappears, but an organism will try to pass as something it's not, if enough of its traits transform. We are nucleate, heterotrophic, vertebral, brainy, bony fish. We are fish.[11]

Three hundred eighty million years ago, give or take, some of us fish made a go of it in shallow water, near land. We were tetrapods. Some of our fins

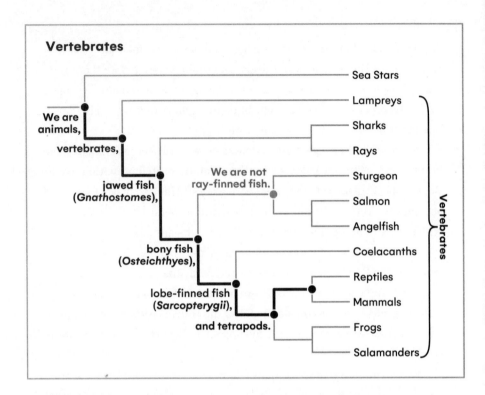

began to seem more like limbs than fins, their bony, muscular extensions became our hands and feet, our fingers and toes. Moving all the way onto land, though, is hard. Terrestrial life takes work, and while land is a vast and promising frontier for those who can do it, the compromises are significant. Everything from holding yourself up, and not being crushed by gravity, to the different ways that light, sound, and odors travel in air compared to water, needed to be dealt with in this new world. Nearly every system needed to be retooled. For a long while, we retained our close relationship with water, lounging in it to keep our skin, our major respiratory organ, functional, and returning to it to breed. Many individuals made mistakes, costly mistakes, deadly mistakes. It all could have turned out so very differently. Our ancestors' errors proved to be survivable or—sometimes— not errors at all, in retrospect. It almost seems preordained, that it would be us discovering our own history, and writing about it, rather than an "evolution played out differently" version of dolphins, or elephants, or parrots discovering and reflecting on their history . . . or farther afield yet, bees, or octopuses, or chanterelle mushrooms.

These early tetrapods, amphibians all, stayed close to water, except when they didn't. Those individuals that ventured far from water took significant risks in doing so, and most of them surely died. All of them were explorers in their way; most of them, like many explorers, took a risk that did not pay off. But those that did not perish found landscapes uninhabited by other vertebrates, and abundant food. So our amphibian ancestors spread across the land, a hot, humid landscape in which the world's first forests were forming, and in many dank corners, giant millipedes and scorpions scuttled and roamed.

Three hundred million years ago, the Earth's continents were all lumped together into the single landmass known as Pangaea, fit together like puzzle pieces. Pangaea was a lush, warm world of abundant plants and giant insects. Even the poles of the planet were free of ice then. Into this world emerged a new egg. The old egg was simple and fragile—it is the egg still used by salmon and salamanders, frogs and flounder. This new egg though, the amniotic egg, had so many protective and nourishing layers that individuals could move their lives farther from fresh water. Finally, we were free of needing such large amounts of water. We were early reptiles; we were amniotes. We are also, still and forever, fish.

Three hundred million years ago, we were on land, with lungs and a fancy new egg. We amniotes evolved from Reptiliomorphs—broadly speaking, reptiles—so all of us amniotes are reptiles, too. Reptiles split and branched, as clades do. Early on in our lives as amniotes, a branching occurred, between the lineage that would further diversify as reptiles, and the one that would become mammals.

Some reptiles lost their teeth and grew shells, and we call them turtles. Some reptiles developed forked tongues and paired penises, and we call most of them lizards. Later, some lizards lost their legs, and some of those legless lizards are what we now call snakes. Even without legs, though, snakes are still tetrapods, because their history doesn't change just because their form has. Some reptiles became dinosaurs, and some dinosaurs became birds. (So, yes: Dinosaurs are not extinct. Birds are dinosaurs. And birds are also fish.)

Birds and mammals have a most recent common ancestor at the base of the reptile tree, and this ancestor of ours was low-slung and slow, cold-blooded and asocial, and didn't have much going on cognitively. Both the lineage that would become birds, and that which would become mammals, independently and without input from the other, evolved into beings that run hot, stand up tall, move fast, and have big, hyperconnected brains. It's a more expensive path through the world, being warm-blooded and big-brained, and birds and mammals have addressed the expense, and its problems, in different ways, but for each of these two groups, it has worked out well for us.

Both birds and mammals have more cultural learning and social com-

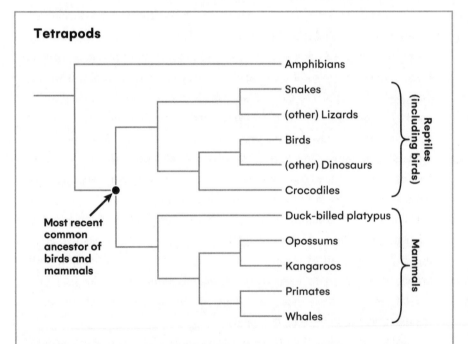

Tetrapods

Amphibians
Snakes
(other) Lizards
Birds
(other) Dinosaurs
Crocodiles

Reptiles (including birds)

Most recent common ancestor of birds and mammals

Duck-billed platypus
Opossums
Kangaroos
Primates
Whales

Mammals

Relationships between tetrapods are depicted here. Among the reptiles, three relationships are worth special note:

1. Snakes are the largest clade of legless lizards.
2. Birds are the only clade of dinosaurs that did not go extinct sixty-five million years ago.
3. Turtles and tortoises ("Testudines") are unambiguously reptiles, but who their closest relatives are remains in question, so they are excluded from this tree.

plexity than any other organisms we know. Running hot and fast, in the particular iteration of history we experienced, seems to have contributed to the evolution of culture. Many species of birds have long lives, long developmental periods, high rates of monogamy, and bonds between individuals that last several seasons, even a lifetime. Some pair-bonded birds duet with each other so tightly that it can be difficult to tell that more than one bird is singing. The same can be said of some pairs of humans.

At the base of the reptile tree, our ancestors branched off, and mammals developed our eponymous trait: mammary glands. Except for a few odd duck-billed platypuses and echidnas at the base of the mammal tree, we mammals have gestation and live birth as well. Parental care, at least from mom, now became unavoidable. Communication between mother and embryo in utero takes many forms—mostly chemical. After birth, some mammal mothers merely provide milk, itself a source of rich immunological, developmental, and nutritional information, but most also protect, and teach, their offspring. Once some parental care was mandated by anatomy and physiology, more was likely to follow.

We are not mammals because we have mammary glands, or fur, or three small bones in our middle ear, though. We are mammals because we have descended, over tens of millions of generations, from the very first mammal, which roamed Earth nearly two hundred million years ago.[12] That first mammal did indeed have mammary glands, and fur, and three small bones in its middle ear. It is those characters,[13] in part, that allow us to diagnose a mammal as such when we find one. But it is our evolutionary history, our ancestry, and the lineage to which we belong that make us mammals, not the characters that we have.

That first mammal was almost certainly small, and nocturnal, and not very bright, by modern mammalian standards. Its fur helped it stay warm, and its ability to lactate provided safe and easy nourishment for babies. With those middle ear bones, it could hear better than its ancestors. It probably had an enhanced sense of smell. The parts of the brain that had been involved in olfaction (smell) for hundreds of millions of years were now expanding and being co-opted into new functions: memory, planning, scenario building.

Our mammalian brains are a collection of small, nimble parts that sometimes act out of view of the others, with both integration and oversight from the larger structure. Our cerebral hemispheres were not always split, as they are now, but lateralization provided the possibility for asymmetric activities on the left and right sides, and a thick band of nerve fibers—the corpus callosum—came to connect the two sides in mammals. Our brains thus illustrate the tension between specialization and the integration of parts.

The first mammal also had a four-chambered heart, which keeps blood that has just been enriched with oxygen in the lungs separate from blood that has been depleted of oxygen from its tour around the body. This allows for a more efficient and capable cardiovascular system. Mammals became endotherms (warm-blooded, generating our heat internally, rather than relying on external sources), evolved new kinds of insulation, and began to experienced REM sleep. (And once again, birds independently evolved all of these traits as well, albeit sometimes in different form—feathers to our fur, for instance.)

Early mammals also solved a problem that had been with us since we came to land. When the earliest tetrapods became terrestrial, their side-to-side locomotion, which salamanders and lizards still exhibit, compressed their lungs, such that moving and breathing at the same time was an impossibility.[14] This side-to-side locomotion put an upper constraint on speed, and on the distance that one could travel before requiring a rest. Anyone who has spent time watching lizards in the wild will recognize the zippy bursts that characterize their movements, followed by rapid breathing. Mammals solved this problem by switching the axis on which we undulate—we undulate up and down, rather than side to side. Now we have the freedom to run and breathe at the same time. It's a useful skill. Add to this another new mammalian feature, the diaphragm, the big muscle below our lungs that coordinates breathing, and mammals can now go faster, and longer, than our forebears. All of this comes with a cost, metabolically, of course; it takes considerably more calories to keep a mammal going than a lizard of the same size.

Now that we were literally running hotter and faster than ever before, our computing power went up, too. Early mammal adaptations allowed for greater efficiency in circulation, respiration, locomotion, and hearing. Early in our mammal history we also became more efficient at chomping on things and in getting rid of waste in the form of urine.[15]

We humans are the beneficiaries of those evolutionary innovations of so many tens of millions of years ago, as are our cats and dogs and horses, as are the squirrels and wombats and wolverines.

There were steps necessary to our becoming what we are, but how many of those would have been necessary for similarly conscious organisms in a rerun of history? What if we could start back at the beginning and try this historical experiment that is *life on Earth* one more time?

In a rerun of history, the chances that the most conscious organisms on the planet would have a four-chambered heart, five digits, and eyes that are built backward are low. But in a rerun of history, in which conscious beings once again emerge, selection would certainly have figured out some way to do an end run around its own inadequacies, creating brains—no matter the specifics—that can look into the future, even when selection itself cannot.

Sixty-five million years ago, the Chicxulub meteor hit the Earth near the Yucatán peninsula. Its impact kicked up so much dust that the sun was blocked for years. Photosynthesis ground to a halt. On the other side of the planet, perhaps accelerated by Chicxulub, one of the largest volcanic features on the planet was forming, the Indian Deccan Traps, belching out large amounts of climate-changing gases.[16] Mass extinctions followed, including that of all the (non-avian) dinosaurs, which had been doing pretty well for themselves for many tens of millions of years.

There is still disagreement about how long it took mammals to begin to diversify, to turn into the great chaotic mess of nearly five thousand mammal species extant on the planet today—half of which are rodents, another quarter of which are bats, and the remaining quarter of which include forms as varied as dolphins and kangaroos, elephant seals and antelope, rhinoceroses and lemurs.

Sometime back when dinosaurs still reigned, primates emerged from the mammalian ranks.[17] Against the odds, our primate ancestors managed to survive the mass extinctions sixty-five million years ago—as did the ancestors of every other organism on the planet today.

One hundred million years ago, well before Chicxulub, the common ancestor of all humans was a small, nocturnal, tree-dwelling primate. It was cute and fuzzy[18] and lived in small family groups. As primates, we developed greater agility, dexterity, and sociality. We primates are still eukaryotes, animals, vertebrates, craniates, bony fish, amniotes, and mammals, each successive, less inclusive group providing greater precision, rather than putting the lie to any earlier group membership. Primates developed opposable thumbs and big toes, acquired pads on our finger and toe tips, and replaced claws with nails. Everything about our hands and feet was becoming more dexterous, more suited to fine motor activities.

We early primates became excellent climbers, too, by virtue of the terminal long bones in our legs and arms becoming less cemented to one another, less stuck in place. Climbing ability came at the cost of some stability on flat ground, which provided even more reason to hang out in trees.

As primates, we became more visual and less olfactory. Our noses shrunk, and our eyes grew. Primates are not as good at the chemical senses— olfaction, taste—as are the other mammals. Just as mammals before us got brainier relative to their ancestors, we primates got brainier, too, compared to the other mammals. At the same time, gestation length expanded—babies cooked for longer inside mom before being born. Litter size fell, so mothers had fewer children at a time to tend to. The period of parental investment after birth lengthened and intensified, and sexual development happened later and later, giving ever more time for young primates to learn how to feel, how to think, how to be.

Monkeys (*sensu lato*), a subset of primates to which we belong, continued these trends. We became almost exclusively diurnal, and even more highly reliant on sight. Our noses shrank further, and our eyes became even larger in our skulls.

Monkeys have singletons or twins rather than litters—and accordingly, all the extra sets of nipples disappeared, the ones that would never be

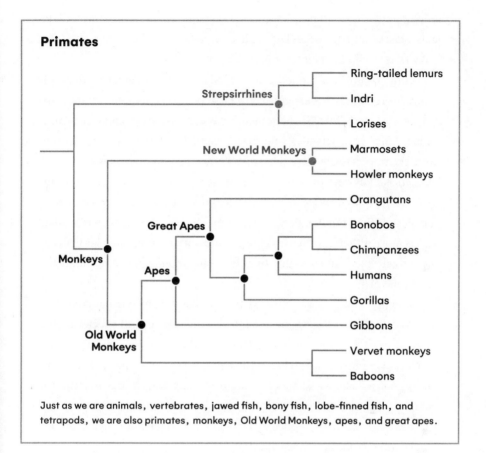

Primates

Just as we are animals, vertebrates, jawed fish, bony fish, lobe-finned fish, and tetrapods, we are also primates, monkeys, Old World Monkeys, apes, and great apes.

needed to feed young . . . except the ones on males. With even fewer babies to tend to at a time, monkey mothers—and far more rarely, monkey fathers—spent more time with each child, teaching it how to be a monkey.

Instead of breeding seasons, during which time every female is fertile, monkeys reproduce on individual cycles. We mate when conditions are right. Humans have overlaid this with the narrative, "We mate when we choose to." There is choice in when and with whom to mate, of course, but there are also underlying conditions that render pregnancy more or less likely to succeed, and they most surely correlate with our feelings of desire and choice, whether we know it or not. Some of these conditions apply population-wide: in time of famine, nearly nobody reproduces, as individuals lack the nutritional and physiological resources with which to bring a baby to term, and feed it after it's born.

Other conditions, though, are particular to the individual: Is your body ready for its first pregnancy? If you've had previous pregnancies, how old is your youngest child? Is she weaned? Do you have older children around to help? Sisters or friends? Your preferred mate? When breeding seasons were the rule, reproductive timing was synced up, so there was lower variance in the answers to these questions. It was also easier, with breeding seasons, for a single male to monopolize the reproductive efforts of several females. With individual cycles, male monopolization of female reproduction is more difficult, which lays the groundwork for relationships between individual males and females to evolve—for monogamy, and biparental care, to evolve.

Twenty-five to thirty million years ago, apes evolved from monkeys,[19] and we are them. Other apes living today include several species of gibbons, by most accounts the most beautiful of the living apes: deeply furred, pair-bonded individuals living in the canopies of the tropical rain forests of Southeast Asia. Some species sing to one another at dawn and dusk, communicating location, certainly, but also, perhaps, information (*This tree has delicious fruit.*), concern (*Do you have the children?*), or intention (*I'm heading home now. See you soon.*).

One of the innovations of apes is brachiation—we swing really well. The cartoon vision of monkeys swinging arm over arm through trees is not nearly so accurate as one of gibbons, or chimps, or even us, doing the same.

The other apes, the so-called great apes, are less beautiful, but even brainier. Orangutans, like gibbons, live in and near the tropical rain forests of Indonesia. Gorillas, chimps, and bonobos are all restricted to sub-Saharan Africa.

More than six million years ago,[20] our ancestors (*Homo*) split from the ancestors of chimps and bonobos (*Pan*), who are our closest relatives living today. It would be millions of years yet before modern humans would evolve, or before modern chimps or bonobos would evolve, either, but the question of what our most recent common ancestor looked like is an intriguing one. One way to approach it is to imagine that it was either more chimp-like or more bonobo-like.

Among those imagining a chimpy past, without recognizing that this is what he was doing, was 17th-century philosopher Thomas Hobbes, who famously declared that humans, in our "state of nature" (that is, without government), are destined to live lives that are "solitary, poor, nasty, brutish, and short."[21] More recently, intellectual luminaries from Sigmund Freud to Steven Pinker have, similarly, imagined that humans need civilization to save us from our basest instincts. It is true that chimps tend toward war rather than peace, and are often found fighting at the edges of their territories.

Bonobos, in comparison, tend toward peace rather than war, and at the edges of their territories, they're more likely to be sharing food with another troop than beating up on each other.

But humans engage in both war and peace. Whether we take up arms when strangers show up at our door, or provide alms and invite them in to share food with us, is highly variable across both cultures and contexts. Given that we are exactly as equally related to both chimps and bonobos, looking to one rather than the other for insight into what humans are makes little sense. We have much to learn from both.

Chimps and bonobos, our closest living relatives, communicate with each other via facial expression and gesture. Their faces don't have the expressiveness that ours do, though—we have more muscle control, and we have whites in our eyes. Their gestures are meaningful and abundant—one chimp can ask another to come with them, give them an object, or move closer. While chimps also vocalize, their utterances cannot, given their laryngeal anatomy, begin to match human linguistic capacity. Gesture and onomatopoeia are firmly rooted in the tangible world; as humans expanded our linguistic arsenal, we were able to explore abstraction more easily.

Humans are long-lived, and have generational overlap, learning not just from parents but also from grandparents. We have large permanent social groupings, culture, complex communication, grief, emotion, and theory of mind. Look to baboons, parrots, chimps, elephants, social dogs, corvids (crows and jays), and dolphins, to start, for these traits in other species. Those organisms are not all closely related, though, so this cluster of human-seeming traits evolved convergently, repeatedly, over history.

Three million years ago, North and South America came together, forming the Isthmus of Panama, closing the connection between the Pacific and the Atlantic Oceans. No hominins were anywhere close to the Western Hemisphere at that time, so wholly unhindered by us, the flora and fauna of the Americas began to interchange, with camelids moving south and ultimately evolving into the llamas and alpacas of the Andes; marsupials moving north, most of which went extinct, with just one small lineage of opossums left to represent marsupials throughout the New World.

Sometime after our ancestors diverged from *Pan* (chimps and bonobos), we moved down out of the trees, having gone into them many tens of millions of years earlier, long before we were even primates. Around the same time that we came out of the trees, our ancestors stood up on two legs, slowly becoming more and more hind limb dominant, losing our prehensile big toes, becoming stable once again on flat ground, shifting the shape of our pelvis and the musculature around it. The landscape in which these ancestors lived was not homogeneous, so standing tall likely brought benefits in both seeing over tall African grasses and breathing while wading in shallow waters.[22] The changing biomechanics of our newly bipedal gait also brought greater efficiency in overland travel, such that bipedalism may well have facilitated multiple new modes of food acquisition: both long-distance hunting and shallow-water fishing.[23]

Gait changes have repeatedly opened up new niches, and therefore new worlds, to us and our ancestors. In the case of bipedalism, our hands also became free to carry things,[24] such as tools. Crows and chimps and dolphins are all known to fashion and use tools, but all are limited by their ability to carry them places. Humans, however, don't just transport tools without them getting in the way of our ability to move through the world; we can even use them while we're on the move (depending on the tool).

Furthermore, standing up on two legs had cascading effects throughout the body, including, ultimately, the restructuring of the human vocal tract, such that we can now create more sounds than any other animal of

similar cognitive ability. It is possible that becoming bipedal was a necessary precursor to having speech.[25]

Two hundred thousand years ago, the bodies and brains of our common ancestor were those of a fully modern human. Take an ancient *Homo
sapiens* from the African Rift Valley, give him a shave and a haircut (and
modern clothing), put him on a crowded, busy street in the 21st century,
and likely nobody gives him a second glance. Except, of course, that he has
no idea what is going on. His hardware is that of a 21st-century human, but
his software is not.

These anatomically modern humans, two hundred thousand years ago,
were hunter-gatherers living in fission-fusion groups on the African savannah, in open woodland habitats, or on the coast. They lived by gathering
plants, by hunting and scavenging wild animals, and, in many places, by
fishing. They were itinerant, never staying in one place for long, although
many would have had regular yearly migrations, returning to particularly
fertile grasslands, for instance, just in time to hunt the grazing mammals—
wildebeest and springbok, among others—which had returned for just the
same reason.

Today, this method of subsistence is restricted to a few human populations, including Mbuti pygmies, !Kung Bushmen, and the Hadza.

The history of early humans is complex, rife with active hypotheses
and interpretations, and reticulate: probably, we split from others like us,
and then occasionally came back together with them to breed. Those
histories—the histories of the Denisovans and the Neanderthals and the
hobbits of Flores island, among others—are better told elsewhere.

One clear trend in humans is this: As early humans collaborated ever
more with one another to gain control over their environment, their biggest competitors soon became each other. We gained ecological dominance through collaboration, which then set us to focusing on competing
with others of our own kind. We cooperate to compete, and our intergroup
competition became ever more elaborate, direct, and continuous, until finally becoming nearly ubiquitous in modern times.[26]

Oscillating between these two challenges—ecological dominance and social competition—we became expert at exploring new niches. We are the ultimate niche switchers.

By forty thousand years ago, many populations of people were engaged in hunting and gathering that was even more cooperative and forward-looking. From that time, the archaeological record begins to show evidence of burial of the dead; personal ornamentation, including the use of skin pigment; and both parietal (2D art carved on rock surfaces) and portable art, including musical instruments.[27] And while archaeological evidence has focused on Eurasia, the ancient cave art from Europe, which has been known for many decades, is no older than more recently discovered art from Indonesia.[28] New discoveries are happening all the time, though, many of which upend our previous assumptions about what makes humans so special: some cave art in Europe from sixty-five thousand years ago has been attributed to Neanderthals rather than *Homo sapiens*.[29]

Seventeen thousand years ago, when the most famous cave art in Europe, at Lascaux, was being created, Beringians had likely become Americans and were spreading across two vast continents.

Ten to twelve thousand years ago, people were beginning to farm.

By nine thousand years ago, permanent settlements were forming; in the Middle East, Jericho may have been Earth's first city.

Eight thousand years ago, at Chobshi, in the Andes of modern Ecuador, people took cover in a shallow cave, and hunted by funneling guinea pigs, rabbits, and porcupines off a short cliff, retrieving the corpses at the bottom, with which they made food and clothing.[30]

By three thousand years ago, much of Earth's landscape had been modified by human activity—by hunter-gatherers, by agriculturalists, and by pastoralists.[31]

Seven hundred years ago, some humans were in Europe, and many of them died of famine. Not too many years later, many more succumbed to the Black Death. Some humans were in China, living under Kublai Khan's rule, whose empire had greater geographic reach than any empire before. Some humans were in Mesoamerica, beyond the reach of Khan, living in the embrace of the Mayan Enlightenment.[32] Across the planet, humans lived in a multitude of cultures, political systems, and social systems. Most populations knew little about life beyond their borders—they knew only of their nearest neighbors. Seven hundred years ago, only a very few people were connecting with others halfway around the world, sharing ideas, food, language. And those few were restricted to the speed of sail and horses, rather than nearly the speed of light.

Humans have retained the vast majority of these innovations from our history, from brains and bone to agriculture and boats. We breathe air and generate heat. We have efficient hearts, which sometimes fail us. We have limbs and hands and feet. We are dexterous, agile, and social. We walk upright, which allows us to carry things long distances.

We have just a few children at a time, and those children learn from their elders, and from each other. Our facial expressions may unite us; our language less so. We use tools to make more complex tools.

We live in groups and have hierarchy. We engage in reciprocity—exchanging both gifts and blows. We cooperate to compete. We have law and leaders, ritual and religious practice. We admire hospitality and generosity. We admire beauty, in nature and in one another. We dance and sing. We play.

Our differences are fascinating, but our similarities make us human.

Now that we understand something of our deep history, and how long it has taken to become human, we can begin to explore modern innovations—and understand more fully the implications of our ancient history and how it shapes our relationship with modernity. We are experiencing changes across the full spectrum of our experience: to our bodies, our diet, our sleep, and so much more. Many of these changes have come so fast and furious that we should not be surprised when they create damage that is difficult to undo.

Timeline

3.5 bya	Life begins
2 bya	We are eukaryotes
600 mya	We are animals
500 mya	We are vertebrates
380 mya	We are tetrapods
300 mya	We are amniotes
200 mya	We are mammals
100 mya	We are primates
25–30 mya	We are apes
6 mya	*Pan* and *Homo* split
200 kya	We are humans
40 kya	We are artists
10–12 kya	We are farmers
9 kya	We are city dwellers
150 ya	We are industrialists

Chapter 3

Ancient Bodies, Modern World

THE SAN BUSHMEN OF SOUTHERN AFRICA, MOST OF WHOM WERE HUNTER- gatherers until just a few decades ago, have little trouble with the kinds of optical illusions that Westerners struggle with. Consider two identical lines, except that they have arrowheads at both ends going in opposite directions. They appear to us to be different lengths, even though they are not. Our eyes fool us, with help from our brains. When asked to do a simple task—assess which line is longer—we tend to fail. The San do not.[1]

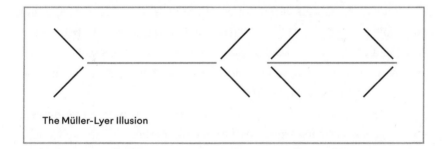

The Müller-Lyer Illusion

Were you to raise an American infant among the San, though, that baby, once it grew up, would not have the same problem her parents did with the optical illusion. Similarly, raise a San child in Manhattan, and susceptibility to the illusion would again show itself. In this case, sensory capability and physiology are being driven by differences in experience and environment—*not* by genetic differences.

Most readers of this book likely live in WEIRD countries: *Western* nations, with a highly *Educated* populace, an *Industrialized* economic base, that are relatively *Rich* and *Democratic*. As societies, we have benefited from industrialization and democracy, which have raised the quality of life for nearly everyone who lives in these countries, but there are many negative, unintended consequences downstream of society-wide changes. While it is clear to most people how much the 21st-century WEIRD environment has expanded the menu of possible experiences that we have at our beck and call, it is less obvious how 21st-century WEIRD life has curtailed other experiences, often to our detriment. Why can we, unlike the San, be fooled by a simple set of lines? It has to do with an alteration in our visual sphere—our homes are clean, climate controlled, and square. Just as depriving kittens of some visual inputs renders them less capable of seeing as adults,[2] perhaps, with our modern comforts and conveniences, we are effectively depriving our WEIRD selves, and rendering ourselves less capable. Or perhaps our visual capacity is being tailored to our uniquely square environment. Either way, modernity is doing something to us at a deeply fundamental level, and the fact that we don't understand it is alarming.

One thing we can be sure of is this: models of human behavior and psychology, which tend to be based on empirical studies of WEIRD undergraduate students, may well be accurate readings of the psychology and behavior of WEIRD undergrads, but they do not inherently make for good models of the rest of the world.

Indeed, it is now clear that we in the WEIRD countries are outliers when it comes to many aspects of the human experience.[3] The implications for this are far more important than being easily tricked by visual illusions, but understanding why we are susceptible to such illusions can provide insight into the risks of hyper-novelty. It is likely that our highly geometric homes and playgrounds, which make up so much of what we see during early childhood, calibrate our eyes such that we suffer from such illusions far more than do those in the rest of the world. That geometry, which we mostly take for granted, emerged in part from being able to run wood through sawmills and create dimensional lumber.

Most people, when their culture began to run wood through sawmills and build homes out of the dimensional lumber that results, would not have thought to ask what, in our human experience and capability, might be affected by this. Dimensional lumber, and the carpentered corners that result, are novel features of the modern human's environment. How has it changed how we perceive the world? Reframing your approach to the world, such that those questions do occur to you, even if you are not sure what the answers might be, is part of the goal of this book.

Compare the change brought on by carpentered corners to the following example of evolutionary change that is understood to be genetic in nature: lactase persistence among adult Europeans.

A majority of adults worldwide cannot comfortably tolerate lactose—milk sugar—in their diet, because they have stopped producing lactase, the enzyme that breaks down lactose. Lactose is a strange sugar, totally unknown other than in mammalian milk. No other mammal species continues to drink milk after weaning. Even among humans, most Asian and Native American people, as well as many Africans, do not do so, so the trait that needs to be explained in humans is not so much the "lactose intolerance" among the majority, but rather "lactase persistence" in that minority of us who continue to enjoy dairy as adults.

The adaptive value of being able to eat dairy into adulthood is varied. Pastoralists of European descent domesticated several species of mammals, and while the value that they received from those animals was diverse—meat, wool, and skins being some of the others—milk was on the list as well. The invention of culinary techniques to preserve dairy for later consumption—in the form of cheese and yogurt, for instance—would have further increased the quantity and frequency of dairy in the adult diet.

Similarly, people at high latitudes gain adaptive advantage from the combination of lactose and calcium in milk. Many of us are familiar with the role that vitamin D is thought to have in facilitating the uptake of calcium—itself critical to bone growth and strength—but vitamin D is rare at the poles. Lactose, it turns out, is a functional substitute for vitamin D in promoting the uptake of calcium. Milk is thus protective against rickets.

Finally, among desert peoples, one of the largest risks is dehydration; being able to digest milk provides both nutritional and hydrational benefits.[4]

So what mechanism explains lactase persistence in some human populations—European pastoralists, Scandinavians, Saharan people such as the Bedouin? The explanation is multipronged, but relatively straightforward: among dairying people and their descendants, a genetic variant that provides a high capacity to digest lactose into adulthood is far more common than among populations that do not regularly consume dairy after weaning.[5]

Raise an ethnically Japanese baby in France, and he is no more likely to be able to enjoy a creamy éclair than if he had been raised back home. Raise an ethnically French baby in Japan, and dairy is a viable option for that child, but probably lacking in availability. Lactase persistence was born of particular environmental conditions, and moved into the genetic layer, where it now lives. It brings differential success in some environments, and not in others, but the experience of being in a part of the world that enjoys or rejects dairy does not affect your ability to digest it.

Following the discovery of the DNA double helix, there emerged a conflation of "evolutionary" traits with "genetic" traits. The terms *evolutionary* and *genetic* began to be used interchangeably, which made it more and more difficult, over time, to talk about evolutionary change that was not genetic. Darwin, had he been aware of Gregor Mendel's work with peas, or had he been around to see the discovery of DNA, would have been pleased to know a mechanism of adaptation by natural selection, but he would *not*, we believe, have assumed that this was the only such mechanism. The conflation of evolutionary with genetic traits became entrenched in popular culture, as in the specious dichotomy of "nature versus nurture." Again, remember the Omega principle (genes and epigenetic phenomena such as culture are inextricably linked, and they evolved together to advance the genes). Asking "Nature or nurture?" isn't wrong simply because the answer is nearly always "both," or because the categories themselves are flawed, but also because once you understand that there is one common evolutionary goal, getting precise about mechanism is less important than understanding why a trait came to be.

The false nature versus nurture dichotomy is disruptive, as it interferes with a more nuanced understanding of what we are and the evolutionary forces that have brought us here. The change in susceptibility to optical illusions seen in WEIRD countries is no less evolutionary than the change in ability to digest dairy in European and Bedouin peoples. The latter has a genetic component, and we have no reason to think that the first one does. Yet they are both equally evolutionary.

If homes full of carpentered corners have made us more susceptible to particular kinds of optical illusions,[6] altering our ability to see, what other costs to a WEIRD lifestyle might exist? As recently as the 1990s, you would have been considered a crackpot if you had suggested that spending your workdays sitting at a desk might have long-term effects on cardiovascular health or risk of type 2 diabetes. No longer.[7]

Carpentered corners create greater susceptibility to certain optical illusions. Overreliance on chairs creates all manner of negative health outcomes. What, then, might deodorants and perfumes have done to our ability to smell the signals emitted by our bodies? What might lives filled with clocks have done to our sense of time? What have airplanes done to our sense of space, or the internet to our sense of competence? What have maps done to our sense of direction, or schools to our sense of family? You get the point.

In this book, we are not arguing for an abandonment of technology. The solution to the many problems laden in a hyper-novel world is not so simple. Instead, we embrace careful application of the Precautionary principle.

When faced with a question of innovation, the Precautionary principle considers the risk of engaging in any particular activity, and recommends caution when the risk is high. In circumstances where the degree of uncertainty about the outcomes of a system is high—when it is not clear what negative effects might result if society engages in, say, carpentering corners or powering our electrical grid with nuclear fission reactors—the Precautionary principle suggests that changes to extant structures should be engaged in slowly, if at all.

Put another way: just because you *can*, doesn't mean you *should*.

Adaptation and Chesterton's Fence

In college, Bret had a friend who was stricken by appendicitis and rushed to the hospital just before her appendix burst. It was traumatic and scary, for her and for her friends. Many of us know of similar stories, and we all know that it's a risk: our appendix might burst, so what the hell are we doing with such a liability in our bodies? Why do we have this famously vestigial organ?

In the early 20th century, medical doctors were wondering the same thing. Many of them came to the conclusion that not only the appendix, but indeed the entire large intestine, was a hazard to humans and that "their removal would be attended with happy results."[8] It was a rare voice that suggested our structures might be adaptive and there might, instead, be a mismatch between our bodies and the quickly accelerating changes brought on by living in a postindustrial culture.[9]

The appendix, we are told now, is vestigial. But *vestigial* is often code for "we don't know what the function is." Has evolution really left us with an organ that is nothing but cost, poses risks to our health, and can be relatively easily surgically removed?

As it turns out, the answer is no. Of course the answer is no.

Many years ago, Bret developed a three-part rubric for establishing whether a trait should be presumed to be an adaptation. It is a conservative test, in that it correctly identifies some traits as adaptations, while leaving undiagnosed as adaptations some other traits that are as well. (In the language of hypothesis testing, the test can therefore produce false negatives—type II errors—but not false positives—type I errors.) This test therefore reveals the sufficient, but not necessary, evidence that something is an adaptation.

Three-part Test of Adaptation

If a trait

1. is complex,
2. has energetic or material costs, which vary between individuals, and
3. has persistence over evolutionary time,

then it is presumed to be an adaptation.

To take movement as an example, swimming takes considerable integration of anatomical, physiological, and neurological systems (among others), and can therefore be understood to be complex; drifting, by comparison—which is definitionally what plankton do—is simple. While the swimming of salmon and the drifting of plankton are likely both adaptive, this test of adaptation does not conclude that planktonic drifting is an adaptation, as it fails the "complexity" component. Many pages could be devoted to defining complexity, variation, and persistence, but take this as a rubric rather than as a quantifiable test.

There are obviously adaptive traits that do not meet the exacting standards of this rubric. For instance, the absence of pigment in a polar bear's fur, and the loss of hair in a naked mole rat, are both cases in which the trait in question involves a savings rather than a cost.[10] When, on the other hand, traits are revealed by this rubric to be adaptations, they presumably are. In combination with the Omega principle, this means that when we see a complex behavioral pattern, like Catholicism, musicality, or humor, we do not need to know the degree to which the trait is based in genes. Even if a trait is transmitted partially or entirely outside the genome, we are logically justified in presuming that its broad purpose is the enhancement of genetic fitness.

Let's try the test on the human appendix.

The appendix, which is found in a smattering of mammals, including some primates, rodents, and rabbits, is an outpouching from the large intestine, and it harbors intestinal microorganisms with which we are in a mutualistic relationship. From us, those intestinal flora get room and board; from them, we get the ability to repel infectious disease, and are facilitated in digestion and the development of our immune system. Furthermore, the appendix is not made of the same material as the surrounding gut—it contains immune tissue.[11] Is it complex? *Check*. It also takes energetic and physical resources to grow and maintain, and has variation in size and capacity both between individuals and between species (*check*). Finally, it has a history, in mammals, that is more than fifty million years old[12] (*check*). ·

The human appendix, therefore, is presumably an adaptation.

Concluding that the appendix is an adaptation, however, does not address the question of what it is an adaptation *for*. The facts that the appendix contains immune tissue and collects mutualistic gut biota are good clues to its function. A recent hypothesis suggests that it is a "safe house" for the gut flora with which we live in mutualism—providing a home for them when we get a gastrointestinal illness and our body purges the pathogens by giving us diarrhea.[13] Diarrhea, as unpleasant as it is, is generally an adaptive response—but it brings with it costs, including dehydration and the loss of your good, mutualistic gut flora. The appendix repopulates the gut with the "good" gut flora after such illness.

Until recently, all humans probably suffered from frequent bouts of such illnesses. The fact that most readers can probably remember, in vivid detail, a bout of such a sickness that left them feeling wrung out and gut-empty is suggestive. We suffer from GI illnesses so rarely that they feel unusual to us. By contrast, diseases that cause diarrhea are common in the non-WEIRD world, and are a significant cause of mortality, especially among children.[14]

More than 5 percent of people in WEIRD countries will have an inflamed appendix at some point in their lives; 50 percent of those will die absent medical intervention.[15] Yet appendicitis is almost unknown in non-industrialized countries, except in areas where Western lifestyles have been adopted.[16] Conversely, in places where diarrhea is still common, appendicitis is far more rare. Perhaps the appendix, which has become a liability for 21st-century people living in industrialized countries, continues to have value for those living with greater exposure to pathogens.

Thus, appendicitis is a disorder of the WEIRD world. So, too, are many allergies and autoimmune disorders, for which there is solid, and growing, evidence to support the "hygiene hypothesis." The hygiene hypothesis posits that because we live in ever-cleaner surroundings, and are therefore exposed to ever fewer microorganisms, our immune systems are inadequately prepared, and so develop regulatory problems, such as allergies, autoimmune disorders, and perhaps even some cancers.[17] Our immune systems are not functioning as they evolved to do, suggests the hygiene hypothesis, because we have cleansed our environments too thoroughly.

Our appendix seems likely to have suffered the same fate as our immune systems. Absent frequent bouts of diarrhea, which are the body's way of ridding itself of pathogenic gut bacteria, the appendix turns from being an important repository of good bacteria to being a liability.

There is an important parable to be invoked here, Chesterton's fence, named for turn-of-the-20th-century philosopher and writer G. K. Chesterton, the man who first described it. Chesterton's fence urges caution in making changes to systems that are not fully understood; it is thus a concept related to the Precautionary principle. Chesterton wrote this of a "fence or gate erected across a road":

> The more modern type of reformer goes gaily up to it and says, "I don't see the use of this; let us clear it away." To which the more intelligent type of reformer will do well to answer: "If you don't see the use of it, I certainly won't let you clear it away. Go away and think. Then, when you can come back and tell me that you do see the use of it, I may allow you to destroy it."[18]

Chesterton wrote this in the same era when some medical doctors had decided that the large intestine was a waste of space in the human body. If Chesterton's fence suggests that a fence should not be removed until you have discovered something of its function, the appendix and large intestine might be called "Chesterton's organs." Keep an eye out for other things that we moderns might be trying to rid ourselves of without sufficiently understanding their function—not only Chesterton's organs, but his gods and his breast milk, his cuisine and his play.

Trade-offs

Chesterton's fence reminds us that things that have been built by humans, or have been selected for over many generations, are likely to have hidden benefits. At the point when early 20th-century doctors were declaring the appendix and large intestine to be not only useless, but actually hazardous, people aware of Darwin, or of trade-offs (or both), should have been able to put on the brakes. Whatever the problem with the large intestine was perceived to be, it would have been smart to try to figure out what benefit it brought before yanking it out and tossing it away.

In everything there are trade-offs. There are hundreds, if not thousands, of distinct competing concerns in any given organism, so how can you possibly know where to begin to look for the trade-off relationships? In fact, pick any two traits, and they are in a trade-off relationship. Trade-offs exist, like the peaks on an adaptive landscape, whether they have been discovered or not.[19]

Broadly speaking, there are two types of trade-offs.[20]

Allocation trade-offs are the most obvious, well studied, and famous. They are the ones that people assume you're talking about if you just say "trade-off." Because many things in biology are zero-sum (meaning that there is a finite amount of resources to pull from—the pie has a size that does not change), we can easily intuit that, if you're a deer, in order to make a larger set of antlers, something else has to give. You've got to borrow from elsewhere in order to get bigger antlers—perhaps you lose some bone density, or spend down other reserves. Under some conditions, perhaps you could just start eating more, and grow more antlers, but this raises the question: If it were that simple, and eating more benefited you in this easy way, what prevented you from doing it before? Assuming that something is constraining your diet such that it cannot simply be increased, growing bigger antlers means having less of something else.

The second type of trade-off is *design constraint*. Unlike allocation trade-offs, design constraint trade-offs are insensitive to supplementation—you cannot just add more of something to solve the problem. For instance, robustness (broadly: being big-boned and muscular) is valuable, as is locomotor efficiency, but you can't maximize both—again, something has to give, and the problem cannot be solved with more resources. Similarly, if you're a bird (or a bat, or an airplane), you can fly with speed, or agility, but you will be only middlingly fast and maneuverable if you try to maximize both. Some other bird will be faster, a third one more agile—but you, you can be a generalist, which is its own kind of success.

The speed versus agility trade-off is easily seen in the body shape of fish.[21] The deep body form of, for instance, angelfish allows them to hover in place and to turn at tight angles while moving almost not at all. This is useful if one of your main jobs in life is nibbling stuff off coral. Compare the shape of an angelfish to that of a sardine, however: long and thin, the sardine is best at speed over straightaways. It can probably pull a good zig in advance of a predator's zag, but it can't stand still.[22]

So design constraint trade-offs reveal that you can't be both fastest and most maneuverable. Design constraint trade-offs also reveal that you can't be both the most robust and the most efficient.

Less intuitively, you also can't be fastest and bluest.[23]

Humans do a great job of evading some trade-offs, of course. For instance, we have addressed the fast-versus-impenetrable trade-off by building outside of ourselves—by having extended phenotypes.[24] The domestication of horses, and advent of castles, allowed some humans to be both fast and impenetrable. And as we discussed in chapter 1, we seem to have beaten the odds when it comes to the trade-off between specialists and generalists.

Humans are a broadly generalist species, with the capacity for individuals—and cultures—to go deep and specialize in myriad contexts and skill sets. Near the Arctic Circle, specializing in seal hunting is a path to survival, but it has no benefit in Omaha, Oxford, or Ouagadougou. Harsh environments tend to require cultural specialization. In less harsh environments, cultures and their populations can thrive by being farsighted as a whole, while encouraging specialization within individuals. At their peak, the Maya had many farmers, but they also had scribes and astronomers, mathematicians and artists. Had any of those artists or astronomers been required to justify their share of the harvest, it might have been hard to do. Those who can see the value of both physical and mental work, perhaps those who have tried both but are likely not best at either—the generalists—are often necessary to reveal the value of one type of specialist to another.

But for all of our cleverness, we can't evade all trade-offs. Presuming that we can is one mistake of Cornucopianism, which imagines a world so full of both resource and human ingenuity that, magically, trade-offs no longer rule (a topic that becomes relevant again in the final chapter of the book). Related to Cornucopianism, or perhaps fueling it, is the fact that the Sucker's Folly can create the illusion that we have conquered trade-offs by blinding us with the richness and opulence of our short-term gains. This is a mirage. The trade-offs are still there, and the cost for all that wealth will be paid, either by those who live elsewhere or by our descendants.

Trade-offs are unavoidable, but this has a remarkable upside: it drives the evolution of diversity. A fine example of this is a set of work-around solutions from plants: Photosynthesis, the process by which plants convert sunlight into sugar, occurs in the vast majority of plants in a form known as C3. C3 works best under conditions that are easy for plants—moderate temperatures and sunlight, and ample water. Because C3 photosynthesis requires that the pores on the leaves—the stomata, which allow intake of carbon dioxide—be open at the same time that sunlight is fueling photosynthesis, C3 photosynthesis comes at the cost of substantial water loss through the stomata. C3 plants, therefore, don't do well where water is limited.

As plants began moving into more marginal environments, such as deserts, C3 photosynthesis posed a particular problem, and two new forms of photosynthesis evolved. One of them is CAM photosynthesis,[25] which allows plants to separate in time when they open their stomata to take in carbon dioxide from when sunlight is fueling their photosynthesis. Having their stomata open at night, when temperatures and therefore evaporative loss are lower, allows CAM plants, like cacti and orchids, to conserve water.

CAM comes with a cost, however, and is more metabolically expensive to accomplish than is C3 photosynthesis. But, in environments where sunlight is plentiful but water is not, CAM wins hands down against C3. Another solution to the problem of water loss is not so much biochemical as morphological. As an organism decreases its surface area to volume ratio, becoming ever more sphere-like, the amount of water lost from its surface is reduced. There is no negotiating with the math: more spherical cacti lose less water than long and lean cacti because they have less surface area relative to their volume from which to lose water. Many plants employ multiple strategies, of course—alternative metabolic pathways, in the form of CAM, and shape changes to reduce water loss.

We will see trade-offs in systems throughout this book—from anatomical and physiological concerns to societal ones—and point out how failing to acknowledge them can be disastrous.

Everyday Costs and Pleasures

Does cheese smell good?

The French have described the aroma spectrum of cheeses as being some distance from "the outhouse."[26] Pungent cheeses are closer; milder cheeses farther away. Yet the proximity of a given cheese to a metaphorical outhouse has little to say about whether it is recommended by the person giving the description. In fact, the most prized cheeses are often the ones with the most spectacular levels of fecal character. So while many cheeses smell something like shit, the positive or negative connotation of the smell is a matter of taste, for which there is famously no accounting.

Or is there?

Of all of our senses, olfaction has been the most difficult to explain. It has proved the most resistant to the reductionism of the lab,[27] and the most confounding of the synthesis sought by theoreticians. Even less well understood is the subjective experience of smell. How a person feels about a particular smell varies widely, some of the difference being arbitrary, but much of it predictable by culture and natal experience. Not only that, but individuals are not even self-consistent over their adult lives. Responses to a given smell vary with context, experience, and sometimes even narrative connotation.

If you are reading this book, you are probably at something of a disadvantage when it comes to understanding the predicament of our ancestors. Chances are, you have never been *really* hungry. We can say nearly the opposite about the vast majority of our ancestors. Most creatures are hungry most of the time. Any population that has more than enough resources will tend to grow until there is no longer any surplus; any population that has too few resources will naturally decline. This implies that populations tend to find and then oscillate around their upper limit, a number known as carrying capacity. So if you were to check in on an ancestor at random, there is a good chance you would find them wanting more food than they have.

The fact that you probably have not been really hungry—indeed, you likely have access to *more* food than would be ideal—makes it hard to intuit just how precious food typically is. We moderns have a hard time imagining the risks worth taking to find more food, the lengths one might reasonably go to protect what they have, and the value that might accompany technological innovations that allow people to stretch the value of food they have already acquired. *A calorie preserved is a calorie found*, one could reasonably argue. A unit of food is even more valuable if you can capture it in times of plenty, and consume it when times are lean.

While we tend to think that the goal of cooking is to make food taste better, much of the world's many culinary traditions have more practical goals—detoxifying foods, amplifying their nutritional value, and protecting them from microbial competitors as we carry them across space or preserve them over time. We salt and smoke meats to ensure that microorganisms that attempt to steal them will die of dehydration. We make fruit preserves with high concentrations of sugar for much the same reason. We pasteurize and freeze perishable vegetables to kill the microbes already on them and to exclude all newcomers. These are not the only techniques at our disposal—many cultures have learned the art of beating microorganisms to the spoiling punch. In effect, we rot foods safely, so they don't get the chance to rot dangerously.

When you get to a milk jug after environmental bacteria have begun to capture the contents, your nose gives you clear directions about what you should do next. Even with considerable nutritional value remaining in that half-full bottle, the potential cost of drinking it exceeds the cost of throwing it away. That's why it stinks. Stinking is nature's way of telling you that you would have to be quite desperate to drink that milk. That observation hints at a risk of using milk from domesticated animals as a food resource. Milk evolved to nourish babies straight from their mothers' mammary glands. As such, milk is full of nutrients. But because milk is meant to be consumed immediately, with little or no contact with the outside world, milk has no defense against environmental bacteria, and we moderns must go to extreme lengths—pasteurization, hermetic sealing followed by refrigeration—just to preserve milk for a week or two. Clearly, an ancestor who needed to preserve milk over a long and unproductive winter would have needed a better solution.

One of those solutions is cheese. By rotting milk carefully, using specially cultivated bacteria and fungi that are not pathogenic to humans, milk can be preserved indefinitely. Cheese is such an elegant solution to the problem that, once made, even a block of cheese that is colonized by bad bacteria on the outside can have a thin layer of its surface removed to reveal the fresh, untainted cheese beneath.

The catch is that humans are programmed to be repulsed by the smell of spoiled milk because, in general, it is a bad idea to consume any substance that has been overrun by microbes. How do we override the ancient wisdom—our noses and brains working in tandem to help us avoid the smell and taste of spoiled milk—in order to profit, metabolically and culinarily, from eating cheese?

If you are born into a culture that has perfected the art of cheese making, being repulsed by *all* spoiled milk is costly. What one needs is a means of distinguishing what's good and what's bad. The fact that something smells a bit like an outhouse is not a sufficient guide.

When Heather was doing research on a small island off the coast of

Madagascar in the '90s—sleeping in a tent, showering in a waterfall—the diet was almost entirely rice. During the middle of one months-long field season, she and her field assistant were shipped a large block of cheese. Practically swooning in anticipation, they prepared a makeshift macaroni and cheese and offered it to the two Malagasy conservation agents with whom they shared the island. The men leaned in to smell what was on offer, then literally recoiled and gagged. There is no history of cheese in Malagasy cuisine.

For us moderns, the fact that a cheese is being sold in a store is a pretty reliable indicator that it isn't going to trigger microbial warfare in our guts. For an ancestor, the behavior of one's kin would provide the same sort of guide. In the end, the proof is in the pudding. If one tries some carefully spoiled dairy product and does not get sick in the following hours and days, then it is safe. Add that discovery to the information one's mouth and digestive apparatus have acquired about the nutritional content. If the value is high—and with cheese it will be—then the smell, whatever it may be, is an indicator of concentrated value. That's true, even if it smells a bit like crap.

We could tell a similar story about "thousand-year-old eggs," sauerkraut, kimchi, or myriad other carefully preserved foodstuffs from around the world.

Here is what we have learned so far: We are all born with basic rules of thumb about what we should and shouldn't eat. A peach smells good. A clam that has been sitting in the sun smells bad. Grilled meat smells good. Carrion smells bad. These rules are an initial guess at the net value of a potential food, but if one stops there, then a lot of nutritious, edible things will be missed, and to a hungry creature—which is almost all of them— that's no small matter. There has evolved, therefore, a secondary system that allows us to remap foods according to empirical information that may be picked up from kin (via culture), or perhaps discovered in hunger-driven desperation (via consciousness). We are constantly remapping foods based on their actual value, rather than on our initial reactions. We may acquire a taste for coffee because it stimulates us, and for beer because it carries the nutrition of bread without the short shelf life.

Were this the end of the story, we could rest easy. As a rule, WEIRD people have plenty of food, we can map and remap our preferences as we like, and one person doesn't need to like what another likes. Taste and preference have become increasingly arbitrary in modern times, as our cultural norms have become generic, global, and market driven.

But that isn't the end of the story. Evolutionary novelty rears its ugly head in the story of smell as well.

Do solvents smell good? Unfortunately, many of them do. They are, of course, famously toxic, so we should clearly remap where they go in our internal models to avoid ingesting them. But that update doesn't go nearly far enough. Most noxious smells from our ancestral world are warnings about interacting with an object—it is best not to come into contact with vomit, for instance, or decaying flesh. The *smell* of vomit, carrion, or a human corpse, however, is not itself a hazard.

This is not the case for many smells to which we are now too often exposed. Not only do many solvents smell pleasant, but the very act of smelling them is itself dangerous. Our long-evolved warning system—if it smells bad, be wary—is unreliable in two ways: (1) many solvents smell good to some people, and (2) smelling them is sufficient to cause physiological harm. Just a few solvents that both smell good to some people and are toxic are acetone, widely used as nail polish remover; toluene, which was used in magic markers until recently and is still in many brands of rubber cement; and gasoline. If we don't train ourselves to avoid breathing when these semipleasant smells are in the air, we actually do damage to ourselves.

Worse still, some truly toxic and otherwise dangerous substances encountered in the modern world have no detectable smell to them at all. Natural gas and propane are gases that have no smell we can detect, and each is capable of concentrating in ways that the tiniest spark—even the electrical arc that accompanies the flick of a light switch—can create a massive explosion. The problem of explosive gases accumulating and being ignited is something that no ancestor had to worry about until recently, and so selection has not built in a natural disgust or alarm response. The danger is so great that postindustrial moderns have cobbled together a solution—tapping into a circuit built for disgust because it effectively gets and keeps our attention. Before propane and natural gas are piped into your home, or delivered to a tank outside of it, they have tert-Butyl mercaptan added to them. This compound gives these otherwise stealthy gases a unique sulfurous smell—like dirty socks or rotten cabbage—that we easily recognize, and with guidance now find alarming.

Consider carbon dioxide (CO_2), which triggers profound alarm as concentrations rise in a confined space. It is not a toxin, but in an environment with a high concentration of CO_2, we will asphyxiate. Our detectors for CO_2 are so ancient and deeply wired that even people with brain damage to the amygdala, such that they never panic under other fear-inducing circumstances, find themselves triggered into panic by high concentrations of CO_2.[28]

By comparison with CO_2, carbon monoxide (CO) is extremely dangerous: it binds to hemoglobin, displaces oxygen, and brings on a quiet sleep from which people do not wake.

Why, then, do we have an internal detector for CO_2, which is dangerous at high densities but nontoxic, but we don't have a detector for carbon monoxide, a deadly toxin?

The answer is tied up in evolutionary novelty. Animals take in oxygen and breathe out CO_2. Our ancestors would have occasionally encountered enclosed spaces that were safe to be in temporarily, but in which the actual act of breathing would make those spaces lethal over time. A detector that causes an animal to become antsy, anxious, and in need of going elsewhere as their cave fills with CO_2 is essential equipment. While it would be great to have a similar detector for carbon monoxide, that need is primarily modern, a consequence of industrial combustion. There is no reason to think that a CO detector would have been harder for natural selection to create, but the value of it is simply too recent to be in our hardware yet.

When Harry Rubin, Bret's maternal grandfather, was a chemical engineer at RCA in the 1940s, he was exposed to substances whose safety to humans was unknown (OSHA did not yet exist). When Harry had to walk through clouds of unspecified gas, he would hold his breath, which earned him a reputation for being fearful. In the Pleistocene, men probably learned courage, and the skills necessary to be productive, in part by mocking one another when they showed cowardice. The hyper-novel world of postindustrial Earth, though, renders this a dangerous strategy. In the Pleistocene, the big risks to a human's continued existence were other people, and the occasional hippopotamus, so the models that they developed with the senses and proclivities that evolution had given them, and with help from their peers, would have been sufficient. When, however, the risks include chemicals never before experienced by another human, the situation is much different. Harry was an adventurer—in his sixties he both learned to ski and climbed Mount Whitney with Bret—but he was also alert to what he did not, and could not, know. He outlived his chemist colleagues, who died earlier than they should have; Harry lived to be ninety-three.

The lesson to be derived from all this is clear. We have been equipped by selection with an ability to smell a wide range of compounds in our environment. We are also born equipped with a rough guide as to what kinds of smells should attract us, and which should leave us repelled. That map is crude and imperfect, matched at best to past environments that do not perfectly mirror our present circumstances.

The human ability to remap our olfactory world according to information gleaned from other people, or from the environment itself, would be sufficient but for the rate at which technological progress has changed our environment. We now regularly create and concentrate deadly things that our ancestors would never have encountered, and that we are therefore unable to detect. Smell is no longer a sufficient early warning system for hazards, because detection and harm are now simultaneous in many cases. As we will see repeatedly, the problem that modern humans face is that, although we are built to deal with novelty, the 21st century is characterized by more than we have seen before. We face novel levels of novelty, and selection simply can't keep up.

The Corrective Lens

- **Become skeptical of novel solutions to ancient problems,** especially when that novelty will be difficult to reverse if you change your mind later. New and audacious technologies—from experimental surgery, to the cessation of human development using hormones, to nuclear fission—may be wonderful and risk-free. But chances are, there are hidden (and not-so-hidden) costs.
- **Recognize the logic of trade-offs, and learn how to work with them.** Division of labor allows human populations to beat trade-offs that individuals cannot. And by specializing in different habitats and niches, the human species beats trade-offs that no single population can.
- **Become someone who recognizes patterns about yourself.** Hack your habits and your physiology. What stimulates you to eat? To exercise? To check social media? Understanding the patterns in your behaviors gives you a better chance of controlling those behaviors.
- **Look out for Chesterton's fence** and invoke the Precautionary principle when messing with ancestral systems. Remember this: "just because you *can*, doesn't mean you *should*."

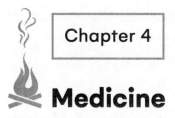

Chapter 4

Medicine

WHEN HEATHER WAS YOUNG, SHE FREQUENTLY GOT STREP THROAT. AS an adult, the strep disappeared, but she began getting laryngitis at least yearly, sometimes a few times a year. It was bad enough that, with some regularity, she thoroughly lost her voice and couldn't lecture. One of those times, in 2009, she gave the following brief presentation—via text on a screen—to her students.

> The medical profession's response to my frequent laryngitis is that I really ought to take some pharmaceuticals, and then some more to counteract the side-effects of the first ones. Why those pharmaceuticals? Because in some cases they reduce some instances of the inflammation that can cause laryngitis. What are the shared symptoms between those cases and mine? The medical professionals don't know. Furthermore, they don't seem to care. Just take the drugs, they advise.
>
> I don't do as they ask.
>
> The treatment of the vast majority of medical complaints with drugs, rather than with actual diagnosis, weakens the ability of the medical system even to do diagnosis. It also pollutes the data stream: who knows who is sick with what, and from what origin, if so many people are on pharmaceuticals with unknown side-effects.

> When I show up on the doorstep of the medical establishment with laryngitis again, they ask me, "Are you on our drugs?" When I tell them no, they abdicate all responsibility. If I won't just follow directions, how can they help me?

Following directions when the people giving them seem to have no idea what they're doing, or why, is neither honorable nor smart. The medical system has been reluctant to take up evolutionary thinking, opting instead for pharmaceutical fixes that too often create new problems, and mask rather than cure the old ones. Anything with a simple biochemical switch would almost certainly have been "solved" by selection by now, *if* it were possible without triggering unacceptable trade-offs, and *if* the "problem" being solved were really a problem.[1]

The modern world is so full of novel inputs that diagnosis has become ever more difficult. Add to this the quick fix that pharmaceuticals promise to provide; the ubiquity of simple and often wrong answers on the internet; and market forces that push health-care professionals into narrower and narrower slices of time with patients, and we should not be surprised that many people feel unseen, forgotten, or dismissed by modern medicine. Chronic illness, constant inexplicable headaches, a vague pain where there ought not be one—too many people come to live with one or more of these irritations, some of which prove to be more than mere irritations. In this chapter, we aim to provide some tools that will help you understand and improve your own health.

Regarding Heather's recurring laryngitis: some years later, with no application of pharmaceuticals, it mostly went away. There never was a diagnosis or explanation.

Against Reductionism

Monarch butterflies raised in captivity don't know how to migrate.[2] Attempt to keep indri—large lemurs that eat dozens of species of leaves on a regular basis[3]—in captivity, and you'll find that you can't replicate their diets sufficiently to keep them alive. Observe a problem (of your own creation) of rats in Hawaii eating your sugarcane crop, attempt to solve it by bringing in mongooses to eat the rats, and find yourself soon thereafter in a landscape bereft of native birds, reptiles, and mammals, but still plenty of rats.

None of these should surprise us—complex systems are just that: complex. Reducing them to a few easily observable, easily measurable parts can feel like success, but reductionism generally comes back to bite those who practice it. Add to this the hyper-novel condition of being able to isolate and synthesize molecules that cause physiological change, and we have a recipe for medicalizing the world, which often makes us less healthy, not more so.

The modern approach to medicine, which can broadly be characterized as reductionist, reveals itself clearly in scientism—an ill-named but important concept. The concept of scientism was introduced by 20th-century economist Friedrich Hayek.[4] He observed that, too often, the methods and language of science are imitated by institutions and systems not engaged in science, such that the resulting efforts are generally not scientific at all. Not only do we see words like *theory* and *analysis* wrapped around distinctly untheoretical and unanalyzed (and often unanalyzable) ideas, but—worse—we see the rise of a kind of fake numeracy, in which anything that can be counted is, and once you have measurement, you tend to forgo all further analysis.

Once we have a proxy for something, a category, we think we know it. This is particularly true if the proxy is quantifiable—if numbers can be attached to it, no matter how flawed those numbers might be. Furthermore, once we have a category, we often stop looking outside of the categories for meaning, as our formal system of carrots and sticks exists solely within the categories.

Calling this "scientism" is a mistake—just like calling the eugenics programs of early and mid-20th-century Europe and America "social Darwinism" is a mistake. Scientism is a bastardization of the tools of science, just as social Darwinism is a bastardization of Darwin's ideas, and a woeful misunderstanding of evolutionary theory.

The mistake of scientism is compounded by the mistake of imagining that we are simply machines, with fixed rules and codes, rather than people. This is the engineer's approach to what humans are (as opposed to the biologist's), and it vastly underappreciates how complex and variable we are. Everyone is susceptible to this error: We look for metrics, and once we find one that is both measurable and relevant to the system we are trying to affect, we mistake it for *the* relevant metric. Calories became *the* metric to keep track of with regard to food, especially for those trying to lose weight, even though calories from carbohydrates, protein, fat, and alcohol have different effects in the body. Pharmaceuticals became *the* preferred treatment for mental disorders—and it is also true that many forms of mental discomfort and distress have been (mis)diagnosed as disorders.

Consider Laura Delano, the woman who was profiled in an excellent 2019 *New Yorker* piece by Rachel Aviv after being overprescribed psychiatric drugs for years. Laura was multitalented and beautiful, privileged by all externally visible metrics, when her internal world began to disintegrate while she was studying at Harvard. Psychiatrists stepped in with a series of diagnoses—including bipolar disorder and borderline personality disorder—and prescribed her over nineteen different psychiatric medications in just a few years. Her doctors saw her drugs as precision instruments—but none of the meds relieved her of her chronic feelings of emptiness and despair. She once even used them to attempt suicide. "I medicated myself as though I were a finely calibrated machine, the most delicate error potentially throwing me off," Laura wrote.[5]

Ultimately, Laura found sufficient resources, both internal and external, to wean herself from the medications, and to see her emotions and moods as fundamentally human, rather than as problems to be solved. While some conditions certainly warrant pharmaceutical intervention, a less reductionist approach to the human body, like the one Laura finally took, involves recognizing what we have been and done for the vast majority of our history.

We are not "finely calibrated machines." We are embodied beings, with feedback systems between brain and body, hormone and mood, that will not be adequately understood or fixed with simple switches. Moving our bodies, as our ancestors always did without needing to think about it, has positive effects on mental health[6]—and is a better first approach to treating mood disorders than are prescriptions. Research on the role that regular exercise could play in improving outcomes for inpatient psychiatric patients is growing rapidly, and results are promising.[7] And while our modern exercise regimens tend to break down our activities into smaller components—for example, cardiovascular, strength training, flexibility—engaging in more ancient activities, be it walking or sports, gardening or hunting, will often integrate all aspects of physical activity without any planning or counting being required.

Furthermore, we are all distinct—what will work for one person may not work for another; this variation between individuals is perhaps the most fundamental of evolutionary observations. Heather used to teach comparative anatomy to undergrads, which comprised ten weeks of dissections of sharks and cats, each student coming to know their specimens well as they studied the visceral, muscle, circulatory, and nervous systems. There were always one or two students who wanted to opt out of the wet labs and instead learn the material from a book or online, but learning anatomy from a book can never replace being in a room with twenty specimens of the same species. Comparative anatomy is nominally about comparisons between species, but comparing individuals *within* a species is, in some ways, even more elucidating. Why, for instance, do the attachment points of muscles never vary between individuals of the same species, but there can be large differences in circulatory anatomy, such that even major vessels like the jugular veins can travel different paths? Because while the function of a muscle changes if it is connected to a different end point, so long as a circulatory vessel gets where it's going, the particular route is not critical. This variation between us as individuals also contributes to the difficulty of predicting whether a solution that worked for one person will work for another.

Considering the Risks of Reductionism as We Choose What to Put in Our Bodies

Is vanillin the same as vanilla? Is THC identical to marijuana? No. In both cases, a single molecule, active and important in the human experience of the larger thing (vanilla, pot), is not representative of the whole. In the case of vanillin, the effect appears to have culinary ramifications only: foods flavored with vanillin do not have the full richness of vanilla. In the case of THC, which has long been understood to be the main psychoactive constituent of cannabis, breeding only for that molecule made plants that would certainly get you high, but they had insufficient antipsychotic tempering effects from CBD, another active molecule in pot. Oops. And as of this writing, there is a new marijuana molecule getting traction in both the scientific literature and the pot-breeding community: CBG.[8] It is being purported to have benefits even greater than those of CBD. Maybe. But it is the human discovery of that molecule that has elevated it to the status of being studied. It was there all along, but now we have imbued it with mystical qualities. Our discovery of it changes nothing about what it does. We often mistake an effect (e.g., of an action, a treatment, a molecule) for our *understanding* of the effect. What a thing does, and what we think (or know) that it does, are not the same thing.

A combination of hubris and technical capacity has humans re-creating this error, over and over again. From fluoridated drinking water to shelf-stable foods with unintended consequences, from the myriad issues with sun exposure, to whether GMOs are safe—we are constantly seduced by reductionist thinking, led astray by the fantasy of simplicity where the truth is complex. Reductionism, particularly with respect to our bodies and minds, is harming us. Sometimes it is even killing us.

Early in the 20th century, fluoride was discovered to be correlated with fewer cavities. So fluoride was put in many municipal water supplies to decrease tooth decay.[9] The fluoride in drinking water is a by-product of industrial processes, though, not a molecular form that appears in nature or has ever been part of our diet. That's one point against it. Furthermore, we find neurotoxicity in children who are exposed to fluoridated drinking water;[10] a correlation between hypothyroidism and fluoridated water;[11] and, in salmon, a loss of the ability to navigate back to their home stream after swimming in fluoridated water.[12] Is fluoride a magic bullet for reducing cavities, with no costs to other aspects of health? Seems not. More to the point, the quest for magic bullets, for simple answers that are universally applicable to all humans in all conditions, is misguided. If it were that easy, selection would almost certainly have found a way. Think you've found a solution that is too good to be true? Look hard for the hidden costs. Remember Chesterton's fence.

The modern food supply chain benefits from shelf stability of processed foods—while the foods along the perimeter of a grocery store tend to be less processed, and therefore less shelf stable, nearly all of the foods in the middle of grocery stores have expiration dates many weeks or even months away. Minimizing fungal growth in our food is certainly desirable, but at what cost? Propionic acid (PPA) inhibits mold growth, and is a prominent additive to processed foods for that reason, but its presence in utero affects fetal brain cells and is linked to an increase in diagnoses of autism spectrum disorder of children thus affected.[13] We should not be surprised that being "shelf stable" comes with costs.

Similarly, people who live near the poles, or who rarely go outside, can come to suffer from short stature, and weak and curved bones—a condition known as rickets. Vitamin D was identified as the missing molecule for such people, and as we moderns seem to like our pills, we are provided vitamin D as a stand-alone product, or as an additive to milk. But what does our history have to say on the subject?

At the end of the first millennium, the Vikings, unlike other northern Europeans, did not succumb to rickets. This turned out to be due to a diet rich in cod. They didn't know that it was the cod that was keeping them healthy and strong, but it was. And we can say with absolute certainty that they did not achieve health through the consumption of a distillate version of vitamin D packaged in pills or tinctures. The historical evidence suggests that most of us could go out in the sun for a bit every day, or eat cod, or do some combination of both, but pills are easier, and they reek of scientism, which is easily mistaken for science and for "taking control of your health." How often have you heard someone say—maybe even said yourself—something like, "I'm being proactive; I'm supplementing with vitamin D!" (Or C, or fish oil, or whatever the quick-fix product du jour is.)

Once again, we should not be surprised to find that this reductionist and ahistorical approach has yielded no evidence to support the claim that adding vitamin D to your diet will keep your bones strong;[14] indeed, lack of vitamin D may be a symptom, rather than a cause, of rickets and related conditions. Not only is loading up on vitamin D not the solution, we are not even sure its scarcity is the problem.

Related to the reductionist thinking around vitamin D is the fact that for decades now, we have received a nearly universal recommendation to slather ourselves with sunscreen whenever we're in the sun.[15] Reduce your exposure to the sun, the logic goes, and skin cancer rates fall. True enough. Guess what goes up when sun exposure goes down, though: blood pressure. And as blood pressure climbs, so do rates of heart disease and stroke. People who avoid the sun have higher overall mortality rates than do people who seek it. A research study on Swedish women reported this remarkable result: "Nonsmokers who avoided sun exposure had a life expectancy similar to smokers in the highest sun exposure group, indicating that avoidance of sun exposure is a risk factor for death of a similar magnitude as smoking."[16] So reductionist scientism has misled us yet again, and likely caused many deaths. Should we stay out of the sun and take vitamin D, or seek moderate sun exposure and get the nutrients we need by seeking something closer to an ancestral diet? An evolutionary analysis suggests the latter. At least on this topic, the medical literature is catching up to that conclusion as well.

Given this track record of reductionist science and health advice, should we trust that GMOs are safe, just because those who would profit from their acceptance, either intellectually or financially, tell us they are? We suggest not. Are some GMOs safe? Almost certainly. Are all GMOs safe? Almost certainly not. How will we know which are which, and can we rely on those who have created them to be vigilant on our behalf? Until we know the answer to those last questions, the Precautionary principle suggests steering clear.

Finally, it is worth noting that some of the major successes of Western medicine—surgery, antibiotics, and vaccines—are firmly rooted in a reductionist tradition and have saved millions of lives. The problem we are highlighting is the overapplication of a reductionist approach. The germ theory of disease—in its simplest formulation, the recognition that pathogens cause much disease—led to the discovery and formulation of antibiotics, a huge health boon for humanity. Then we overgeneralized, and imagined that all microbes are bad for us.

We are now coming to realize that our microbiome has evolved with us and is necessary for a healthy gastrointestinal tract. Antibiotics are one of relatively few powerhouse tools of Western medicine, but as they have been overprescribed, we have seen an attendant rise in people becoming sick, often chronically so. Just as people are falling ill from lacking healthy microbiomes due to overprescription of antibiotics, so too are our livestock. Furthermore, there are unintended side effects of many antibiotics that will be shocking to most people. Heather's personal experience with the unintended consequences of antibiotics was a ruptured Achilles tendon. It is now understood that tendon and ligament injury is one side effect of Cipro (and all of the antibiotics in that class, the fluoroquinolones),[17] which Heather took in quantity in the 1990s to ward off GI bugs while conducting research in the tropics.

From fluoridated drinking water to antifungals in shelf-stable food, from sunscreen to the overuse of antibiotics—over and over we make the same kinds of mistakes. Combine reductionism with a tendency to overgeneralize, in a hyper-novel world where quick but expensive and potentially dangerous fixes are common, and we have explained some of the major errors of modern health and medicine.

Bringing Evolution Back to Medicine

Evolution is the central unifying theory of biology. The implications of this can be subtle, though—subtle enough to confound whole fields, including, to a large extent, medicine.

Ernst Mayr, one of the 20th century's great evolutionary biologists, formalized the distinction between proximate and ultimate levels of explanation.[18] In attempting to tease apart cause and effect in biology, he distinguishes two branches within biology, which many scientists themselves may not be aware of.

Functional biology, Mayr argues, is concerned with *how* questions: How does an organ function, or a gene, or a wing? The answers to these are proximate levels of explanation.

Evolutionary biology, in contrast, is concerned with *why* questions: Why does an organ persist, why is a gene in this organism but not that one, why is the swallow's wing shaped the way that it is? The answers to these are ultimate levels of explanation.

Good science needs both approaches, and indeed, all scientists who consider complex adaptive systems need to have facility in both realms.

Because *how* questions—that is, proximate levels of analysis, questions of mechanism—are more easily pointed to, observed, and quantified than the underlying question of *why*, mechanism has become most of what is studied in science and medicine. Not coincidentally, *how* questions also tend to be what are reported, in breathless sound bites, by the media. Too often, these proximate questions are imagined to be *the* level at which the scientific conversation needs to be had. This serves nobody—not those who are interested in the study of *why*, nor those who are interested in the study of *how*. Some traits are still beyond the scope of our understanding from a mechanistic perspective, but that does not render them immune to analysis at the ultimate level. Even if we don't yet have a sense, for instance, of *how* love or war emerges, that need not preclude us from investigating *why* they do.

"Nothing in biology makes sense except in the light of evolution," said biologist Theodosius Dobzhansky in 1963.[19] Medicine is biological at its heart. That does not mean, though, that most medical research being done is evolutionary in its thinking, or in the questions being asked.

Combine a tendency to engage only proximate questions, with a bias toward reductionism, and you end up with medicine that has blinders on. The view is narrow. Even the great victories of Western medicine—surgery, antibiotics, and vaccines—have been overextrapolated, applied in many cases where they shouldn't be. When all you have is a knife, a pill, and a shot, the whole world looks as though it would benefit from being cut and medicated.

Even bone setting is due for reinvestigation with an evolutionary lens. Bone and soft tissue both respond to force, to being used, in order to grow strong; they are antifragile. When a modern human breaks a bone, though, if it's a long bone—an arm or a leg—full immobilization in a cast for six weeks has long been the prescription. After six weeks of such complete protection, how likely is the bone to have been rebroken? Not at all. How likely, though, are it and the tissue around it to be weak, unprepared for the world? Quite. In this regard, bones and children may be alike. If, instead of coddling your bones, you carefully expose them to the world, not just before but also after a trauma, we argue that they can (in certain situations) heal faster and allow you to get back to your life more quickly.

On Christmas Day 2017, Bret broke his wrist attempting to ride the hoverboard that Toby, our younger son, had given Heather as a gift. Instead of going to the ER, he spent one night in excruciating pain, a second in fairly bad pain, and the next week trying to avoid shaking people's hands at a conference where we were meeting new people. It was socially awkward but, after the first few days, not physically so. He never got a cast, and within two weeks he had nearly complete mobility and strength back. Four weeks after the break, he was as good as new.

A year and a half later, Toby, by then thirteen, fell off a high rope swing on the last day of sleepaway camp. In protecting his head and neck in the fall, he broke his arm. The camp was in the Trinity Alps of far Northern California, and we got him to a fine ER in Ashland, Oregon, where the doctors confirmed the break with an X-ray, gave him a temporary splint, and urged us to get an orthopedic follow-up when we got back home to Portland, at which point a cast would be put on his arm. Our family wasn't going back to Portland for several days, though. Toby spent that first night in his splint, in substantial pain, despite some pain medication. The next day the four of us went on a five-mile hike outside of Ashland; he wore his splint but asked if he could take it off when we were back at the cabin we had rented. His hand and arm were swollen, but he could begin to move his fingers when the splint was off, twenty-four hours after the break. By three days later, the pain was mostly gone, he was off all pain medication, and he climbed, one-armed, up a tall rope structure in Ashland's beautiful Lithia Park. We did take him for an orthopedic follow-up when we got back to Portland, and the good doctor did allow that the splint was an acceptable alternative to a cast, so long as Toby wore the splint all the time, except for showers. We let him stop wearing it altogether seven days after the break. Two weeks after the break, he went for a bike ride. Six weeks after the break, his final ortho check earned him not only a clean bill of health, but also some astonishment from the medical staff at how strong and capable his arm was.

A proximate approach to broken bones identifies the acute problem and arranges a quick fix of the problem. The bone is broken? Cast it! An ultimate approach considers what must have been the case when our ancestors on the savannah broke bones. Some of them died—of infection, of exposure, of being eaten by carnivores. Some didn't, though, and those who didn't would have used pain as their guide to what was possible, pushed their activity to that limit, and no further. Muting pain with medication interferes with the feedback system in our bodies, making it much harder for us to know what we should, or should not, do. Similarly, eradicating swelling in the wake of an injury means that you are much more likely to injure yourself again, in the same place. Swelling after an injury is uncomfortable and cumbersome, and often adaptive; it immobilizes a limb like a dynamic cast. If you let your body communicate with you—with pain, swelling, heat, more—you are far more likely to get back in the game, whatever your game is, sooner, and more safely.

The story that we tell in chapter 9 about the time that Zachary, our older son, broke his arm and required surgical intervention to fix it further reveals the dangers of reductionist thinking, even if that thinking is evolutionary in nature. If we had acted as if all fractures are the same, and time and natural processes would be sufficient to heal them all, our older son would now be in very poor shape. Employing evolutionary logic is not exclusively about discovering our strengths; it is also about understanding our weaknesses and when to augment with modern solutions.

Whom to Believe in the Era of Reductionism and Hyper-Novelty

In this chapter, we have critiqued the reductionism that pervades much of modern medicine. Combined with this is the hyper-novel world in which we live, so complex, so filled with choice and authorities of varying credentials arguing opposite things, that many of us crave simple, immovable rules with which to navigate our lives. We want, at least in some realms, to be able to "set and forget"—to rely on culture, rather than consciousness. This is part of what drives brand loyalty, taking the same commute even when a better one is available, and sticking to pharmaceutical and dietary recommendations even after they've been debunked.

In our quest for set-and-forget rubrics, we fall prey to reductionist thinking. What we need instead is flexible, logic-based, evolutionary thinking with which to navigate. In February 2020, early in the COVID-19 pandemic, both the World Health Organization (WHO) and the US surgeon general repeatedly told the public that "masks aren't helpful" in protecting against SARS-CoV-2.[20] In this case, too many people listened to the authorities rather than thinking through the logic themselves. Why, for instance, if masks are pointless, are they exactly the equipment used by health professionals when trying to avoid infection from respiratory ailments? When the directives were later reversed, people who had followed them based on authority alone lost faith in those same authorities. It was then difficult to regain the public's trust sufficiently to encourage a careful, nuanced approach to reducing the spread and impact of this novel coronavirus. Simple prescriptions make snappier sound bites, and they are easier to remember for those looking for set-and-forget solutions, but when they fail you, you are left with nowhere to stand, no ability to problem solve for yourself. Rather than blindly "trusting the science" or following the lead of authorities, learn to do at least some of the logic for yourself, and seek authorities who are willing to both show you how they arrived at their conclusions and admit when they have made mistakes. Again, our hope is that we can help you become a better problem solver.

How we moderns see our bodies informs how we see our food. Everything from diet culture to eating disorders operates on the reductionist idea that our bodies are machines and can be engineered into compliance. When we look at other cultures, we see a less engineered approach, and more reliance on myth and tradition, which rarely attempt a dispassionate analysis of why they prescribe what they do. The lack of antibiotics in much of the non-WEIRD world has needlessly led to the death of many; we argue that a decrease in reliance on tradition and self-sufficiency in much of the WEIRD world has also led to many deaths. Both can be true. In the next chapter, we will explore some of human history and prehistory, our traditions and innovations, with a focus on food.

The Corrective Lens

- **Listen to your body**, remembering that pain evolved to protect you. Pain is information about the environment, and how your body is responding to it. Some injuries require professional treatment, but some can be monitored without intervention. Pain is both unpleasant and adaptive; think twice before shutting down its message.
- **Move your body every day.**[21] Take walks. Mix it up—don't do the same thing all the time, and don't move your body in the same way whenever you move it. And, at least sometimes, move intensely, and move outside, where the stakes are higher.
- **Spend time in nature, the less constructed and controlled the better.** This has many benefits, among them the dawning recognition that you cannot control everything in your life, and that experiencing discomfort—even the slight discomforts of a too-warm day, or rain for which you are unprepared—calibrates your appreciation for other aspects of your life.
- **Be barefoot as often as possible.** Calluses are nature's shoes, and they do a far better job of transmitting tactile information to your brain than do shoes.[22]
- **Resist pharmaceutical solutions for medical problems if you can.** While antidepressants, antianxiety meds, and more improve some people's lives, they are often not the best solution. Usually, there

are alternatives available; many mood disorders, such as depression, are beginning to be understood by Western medicine to be treatable with diet, ample sleep, and regular activity.[23]

- **Look out for mismatch diseases**, such as adult-onset diabetes, atherosclerosis, and gout.[24] These are diseases that reflect an inconsistency between (one of) your Environments of Evolutionary Adaptedness and your current life. They also tend to reflect affluence, compared to your evolutionary past. For at least some of these, bringing your modern behavior closer to that seen in an older EEA could help mitigate the damage.

- **Consider this informal test to assess certain types of ailments, and whether a modern "fix" is called for:** In environments similar to the one I am living in, did people suffer from this ailment prior to modern medicine? If yes, a novel solution is warranted. If no, look to history for the solution. Take rickets as an example, for someone of European heritage living in the Pacific Northwest. Did people suffer from rickets in such northern latitudes in the past? One type of answer is that evidence suggests that at least some populations of northern Europeans did not suffer from this condition. Seek answers there (remember the Vikings and their cod). A second type of answer is that native people in the Pacific Northwest did not suffer from rickets. What worked for them might not work for someone who is not of native heritage, but it well might. Look to geographically local history for solutions.

Chapter 5

Food

WHAT IS THE BEST DIET FOR HUMANS?

People have been preoccupied with this question for a long time—especially WEIRD people. Many of us have tried diets that are supposed to be "what our ancestors ate." But the lens with which we do this tends to be reductionist and a-evolutionary at best.

From diets designed to alter the pH of your body, to those based on your blood type, to those that restrict your intake to one or a few kinds of food (e.g., grapefruit or cabbage soup), WEIRD people are both obsessed with, and confused by, the question of what to eat. Let's take just two diets that are popular in some circles—two that seem less crazy than many: the raw diet and paleo.

Those who advocate for a raw diet suggest that it is the healthiest, "most natural" way to eat. Cooking, they say, is a modern bastardization of the human diet. This is simply wrong. Not only is cooking ancient in the human lineage, it also allows us to get more calories from food.[1] And while it may be true that cooking can reduce some of the vitamins in the food that has been cooked, the benefits far outweigh this small cost. People on entirely raw food diets are often undernourished, especially if those diets are also vegan. They are generally thin, but that thinness is not inherently healthy.[2]

Others argue for the health of the so-called paleo diet: a diet free of grains and most carbohydrates, and high in fat. This may well be a healthy diet for some people. But those who come from lineages whose cuisine is rich in carbohydrates—people from the northern Mediterranean, for instance—may not be best served or most healthy on such a diet. Further-more, there is growing evidence that early humans were eating a diet rich in carbohydrates from starchy underground vegetables—relatives of which include the African wild potato—as much as 170,000 years ago.[3] This sug-gests that, while healthy for some, the "paleo diet" is not particularly re-flective of paleo ways of life.

These are only two of today's many modern approaches to diet, but they reveal two similarly misguided assumptions about food. First, they imply that there is a fixed and universal answer to the question of what one should eat. Just as we discussed with regard to medicine, the chances of this being true are vanishingly small. Differences in individual develop-ment will render some foods healthy for one person, less so for her neigh-bor. Demographics such as what sex you are will affect what food is best for you, and the simple act of aging will change the answer as well. Cultural differences, which are often based in geography, may well affect your op-timal diet. And those cultural differences may have moved into the gene layer, reflecting population-level genetic predispositions to particular foods—as with the lactase persistence of European pastoralists and Saha-ran Bedouin. Again, remember the Omega principle, which posits that ex-pensive and long-lasting cultural traits like cuisine should be presumed to be adaptive, and that adaptive elements of culture are not independent of genes.

The second misguided assumption that many such diets reveal is that they seem to presume that food is merely for survival. The evolutionary truth is that food is for more than just survival. Food is more than nutrients, vitamins, and calories. Like all animals, indeed all heterotrophs, we eat to acquire the energy and nutrients necessary to be alive. But the human relationship with food, like that with sex, has expanded beyond its original purpose. Humans no longer eat merely to satisfy energetic requirements, any more than we have sex just to make babies.

An ahistorical, reductionist approach to diet attempts to replace food with its component parts. Take this supplement, eat that bar, drink the contents of that can. You'll get X grams of protein, a handful of alphabeti-cally named vitamins, and that rush of energy you've come to expect to get you through your day. As is so often the case, such an approach creates hyper-novelty, which then creates new problems of its own—problems that we are too often defenseless against.

The mistakes inherent in this approach are many, and the hubris abundant. The 20th century saw the dismantling of Chesterton's cuisine. As suggested by Chesterton's fence, we ought to have understood what cuisine was for before we took it apart. In its place, we got easily quantified and commodified pieces and parts that can be added and subtracted on a whim by the producers of processed food. Instead of chasing the newest diet advice with processed food—now with more B_{12}!—we should be eating real food. Real food is that in which the base ingredients are recognizable as coming from a living organism (there are just a few exceptions, like salt).

Some things taste delicious and flavorful to everyone: "rich and succulent" and "salty and crispy" and "sweet and smooth" are combinations that are beloved across cultures. Our sense of taste evolved in an era when meat and other fatty foods, salt, and sugar were all rare. Our sense of flavor is thus evolved, and important. This is true, and it is true that our sense of flavor can be and is gamed in a system that can easily create fat, salt, and sugar, and add them to any foodstuff it wants—another manifestation of hyper-novelty.

Fast food tastes good to many people because it successfully games our sense of taste, accessing single notes—fatty, salty, sweet—in a reliable, uniform way that can be triggered anytime you order the same thing at any one of hundreds of identical stores. By contrast, a plate of carne asada, rice, and beans, with freshly made tortillas, pico de gallo, guacamole, and pickled vegetables from a local taco stand or from your own kitchen, will always be more nutritious (and more delicious, too, for many of us, and for anyone who chooses to develop a palate that finds it so). That plate of less processed, more species-diverse food is more nutritious than a plate of fast food, just as it is more nutritious than taking pills that are supposedly a match for all of the nutritional benefits you're getting from your food. The whole is greater than the sum of its parts.

But why is the whole greater than the sum of its parts or, to put it another way, why is a holistic approach often better than a reductionist one? Two reasons. First, the parts of a given system that we have turned into pills are usually not descriptive of the whole system—remember the discussion of vanillin (a component of vanilla) and THC (a component of marijuana) from the previous chapter. Second, there is often emergence in the combination of food in its less processed form, such that our bodies can use food more effectively than it can use pills. This is especially true for those foods that have a long culinary history together—such as the "three sisters" of corn, beans, and squash traditionally eaten by Mesoamerican peoples. When these foods are eaten together, they constitute a complete protein. Such a long culinary history points to the human discovery, usually unconscious, that just as "smells good" was a good proxy for "good for you" until recently, so too was "tastes good" a good proxy for "good for you."

Reductionism in our approach to food fails us, as our bodies are not static, simple systems, nor do all individuals have the same needs.

There is no universally best diet for humans. There can't be.

In our varied Environments of Evolutionary Adaptedness, there were a few staples—in the Andes, quinoa and potatoes were generally on the menu; in the fertile crescent of Mesopotamia, wheat[4] and olives[5] were among the foods domesticated early; in sub-Saharan Africa, sorghum and guinea yams were significant early agricultural successes.[6] There was meat, sometimes, in short-lived abundance. There was fruit, seasonally, also in abundance. There was alcohol, intermittently, and botanically created stimulants, in some places. In those places, those stimulants were a regular but low-key part of life. Even the ratio of macronutrients is not stable between cultures—Inuit have a high-fat, high-protein diet, with almost no carbohydrates, which is unlike almost any diet that evolved closer to the equator. Given such variation, the idea of a universally best human diet seems patently absurd.

In the 21st century, there are many foodstuffs available that will trick you into eating them, even when many parts of you are sensing that it's a bad idea to do so. Before the advent of cheap, always available highly processed food, our ancient aesthetic preferences made a very good guide to what to eat. Those ancient aesthetic preferences aren't so reliable now.

Hyper-novelty has gamed our ancient rubrics about what to eat and what not to, and thus, we must use our consciousness to separate the good from the bad.

Reductionism in our approach to food also fails us in that it ignores food's ability to provide connection to other humans—with the family and friends who cooked for or with you, or for whom you cooked. A reductionist, nutrient-centric approach to food fails to allow for celebration, or for grief, both of which are often accomplished through food. It fails to recognize and remember cultural tradition, and to consider flavors that have come together through serendipity and experimentation. Cuisines old and new reflect both their terroir—the land from which they emerged—and their borrowing from other cultures and places. Those three sisters of corn, beans, and squash are still dominant in Mexican cuisine; limes, garlic, and cheese, all introduced by the Spanish to people in the New World, have been incorporated deliciously as well.

Humans do not just need protein and potassium and vitamin C. We generally need those things in the food context in which our ancestors ate them. We also need culture and connection. When we sit down to eat a meal together, especially when we are breaking bread that we have ourselves made, we gain far more than calories.

Let us now look to our evolutionary history—how we ate and what we ate, as a way to understand how best we might feed ourselves today.

Tools, Fire, and Cooking

Since long before we split from our chimpanzee-like ancestors, we have been using tools to extract food from our environment. Modern chimps—which are not the same organisms from which we split some six million years ago—show considerable evidence of tool use for food extraction. Some use stones to crack open nuts.[7] At Gombe National Park, Jane Goodall first observed chimps engaging in "ant-dipping," in which they dip a stick into an ant nest, pull it out covered in ants, and lick the ants off.[8]

Both modern chimps and humans adore and pursue honey as well, and sometimes use similar means to get it—by inserting a stick in a crevice and licking the honey-covered stick after it emerges. Among the Hadza, a hunter-gatherer people of East Africa, though, the most successful hunters use two additional tools, and procure far more honey than any chimps can. The first is an ax, which affords them greater precision in reaching the honey; the second is a fire stick, with which they smoke out the bees, making access to the honey far less dangerous.[9]

After our split from a chimp-like ancestor, over six million years ago, our tool-making abilities began to flourish and diversify. By 3.3 million years ago, our ancestors were using stone tools.[10] Two and a half million years ago, our ancestors were using stone tools to butcher the carcasses of animals they had hunted or scavenged, and to extract marrow from the bones.[11]

Our ancestors may have been controlling fire for more than a million and a half years.[12] Fire brings many advantages, of course. It provides warmth and light, a warning and protection against dangerous animals, and a beacon to friends. A little later on in our relationship with controlled fire, we began to use it to boil water and make it potable, to eradicate pests, to dry our clothes, to temper metal with which to make tools. With fire we can see one another and our work at night, and we may gather around it, telling stories or making music. There are no known human cultures without fire, although early reports from anthropologists, missionaries, and explorers often made claims to the contrary.[13] None other than Darwin suggested that the art of making fire was "probably the greatest [discovery], excepting language, ever made by man."[14]

While Darwin didn't expand on his pronouncement, primatologist Richard Wrangham has formalized a related hypothesis: the control of fire, and the subsequent invention of cooking, have been pivotal in making us into the humans that we are today.[15] One of the many advantages of cooking is that it makes our food safer by reducing risks from parasites and pathogens. Cooking also detoxifies some plants, making foods available that would otherwise be inedible to us.[16] It reduces spoilage, such that we can store food for longer, and allows us to open and mash otherwise impenetrable foods.

But none of these advantages, as substantial as they are, compares to this: Cooking increases the amount of energy that our bodies can obtain from food. To obtain sufficient calories from a raw food diet like that of our extant wild ape relatives, humans would have to chew for five hours every day. Cooked food is thus an economical and efficient use of hard-earned food resources, freeing up both time and energy for other things.[17]

Many indigenous cultures have myths about how they started to use fire; fewer incorporate the origins of cooking. We find one origin story of cooking among the inhabitants of Fakaofo, in Polynesia. The story goes that a man, Talangi, approached an old blind lady, Mafuike, and asked that she share fire with him. Mafuike was reticent, until Talangi threatened her, but it was not only fire that he wanted. He also demanded from her the information of which fish were to be cooked with fire, and which were to be eaten raw. This, the story goes, began the time of cooking food.[18]

Just as every known human society has had fire, we have all cooked our food as well.[19] We might well posit that once we were cooking, and therefore pulling more calories from every morsel hunted or gathered, we had more time for other pursuits: for telling stories while preparing food together and, especially, while sitting around the communal fire and eating together. Fire and cooking were necessary precursors to our use of food as social lubricant, as facilitator of culture and connection.

The control of fire, then, can be viewed as an amplifier of conscious exploration. Fire brought people together to dream, to imagine new ways of being, and to collaborate in moving imagined possibility into reality. Our human use of fire unlocked many superior modes of competing: It provided one of several routes to sanitizing and preserving food—which would have allowed for survival during famines, or during long journeys. When those journeys were across water, they could have been facilitated by fire, in part because, under many circumstances, burning out the hulls of trees to make them into canoes is a faster route to having a functional boat than is carving. Being able to take fire with us also opened up territory that was colder than anything we could have survived without it, allowing for exploration of the entire globe.

And the control of fire led to the invention of cooking, which saved both time and energy, and ultimately led to the proliferation of cuisines and methods that we have today.

Persuading Wild Foods to Team Up with Us

We tamed fire, and it was difficult, to be sure, but that activity differed substantially from our taming, or domesticating, of food. Fire is unlike food in that it is wholly indifferent to people. All food—with the exception of a few culinary minerals, like salt—is organic. Food is biotic. It evolved. Therefore, food has—or had, while it lived—interests of its own. Fire, being abiotic, has no interests, no goals. It has never been alive.

Of the food that we eat, which of it has an interest in being eaten? That is, what foods were produced by the organism with the expectation that their product would be eaten?

Milk, fruit, and nectar. That's all.

Milk is produced by mammal mothers to feed their young. Fruit is a plant's way of enticing animals to disperse its seeds—blackberry bushes attract birds and deer and rabbits, and the blackberry bushes achieve their evolutionary goals when those animals feast on berries, wander off, and poop out the seeds, now rich with fertilizer. And nectar is a plant's way of encouraging pollination—blueberry bushes lure bees of many species with the promise of a sweet reward, and the plant achieves its evolutionary goal of reproducing when a bee carries pollen from one flower to another.

Seeds don't want to be eaten.[20] Leaves don't want to be eaten. And certainly, food that requires that we kill entire organisms to eat it—the flesh of animals, be it cow, salmon, or crab—does not want to be eaten.

Yet over millennia, we have persuaded many wild foods to team up with us. Those that are persuadable may become susceptible to horticulture, agriculture, animal husbandry. We are, in some cases, coevolving with them. While their fate hinges more tightly on ours than ours does on most of theirs, we do have shared fate.

Corn, potatoes, and wheat are far more widespread and abundant, and at less risk of blinking out of existence, than they would be absent the human use of them as food. These plant species have, therefore, benefited from their association with us. While it seems a more difficult conclusion to arrive at, for emotional reasons, the same can be said of domesticated cattle, pigs, and chickens. We have increased their range and numbers, and decreased their risk of extinction, by domesticating them for food. In the Salish Sea, off British Columbia, clams grew in size and abundance under indigenous gardening practices, even though those gardens came with direct costs to the clams. Gardened clams were harvested, and eaten, in far greater numbers than were their wild cousins.[21] As we look around the plains of North America and see precious few wild buffalo but many, many cows, it is difficult to argue that buffalo are doing better than cows.

Domesticated species, on balance, therefore have a good evolutionary deal. It will be easy to object to this conclusion, on the basis that a chicken, for instance, is no longer around to enjoy any sort of deal at all after it has been eaten. Consider the population of chickens from which that now-dead chicken came, however: other members of the population are still around, and thriving, far more so than they would be had a long-ago chicken ancestor been more resistant to teaming up with humans.

Extending the evolutionary logic further, we predict that cultivated organisms will gain an adaptive advantage by taking on traits that benefit us. In other words, selection should favor traits in cultivated species that benefit humans, even if the humans that cultivate them are oblivious to it. Food for thought, if you are skeptical of the mutualistic relationship between humans and the organisms we have teamed up with.

Loaves and Fishes

One of the most famous stories from the New Testament is the multiplication, by Jesus, of just five loaves and two fishes into enough food to feed five thousand people. The fact that this story is repeated in all four gospels is enough to give us pause. The miracle attributed to Jesus is the usual focus of people considering this story, but what of the choice of foodstuffs for the army to eat? Perhaps "loaves and fishes" has meaning that is deeper than we might have imagined.

Agriculture has been invented by humans several times around the world, independently, beginning around twelve thousand years ago.[22] Bread, one would think, would follow on the heels of agriculture, a savvy way of preserving and transporting the nutrients of the newly domesticated grains. Yet loaves predate agriculture in at least one culture, and by a substantial margin. In what is now Jordan, the ancient Natufians were making and eating bread at least four thousand years before they were farming.[23] From wild seeds that were precursors to modern wheat (einkorn), and from the roots of tubers, the Natufians made flour, which they then cooked into flatbreads, perhaps in preparation for travel. Flatbreads have the comparative advantages over raw seeds and tubers of being lightweight, highly nutritious, and transportable, as well as having a long shelf life.

Agriculture affords myriad advantages over the harvesting of wild plants. Farmers are now in greater control of both space and time: they know where to go to find the majority of their food and can coordinate when it is likely to be ready for the harvest. Domestication of farmed species further allows for humans to select for those things that we value (e.g., larger fruits, higher fat content, ease of access to the desirable part of the plant), and select against those things that we don't, such as toxins the plant has put on board to keep organisms from eating it.

But we were recognizably human, even culturally so, long before we were agriculturalists. Cooking, which long predated farming in every human society,[24] would have been facilitated by having vessels in which to cook. In China, pottery existed for ten thousand years before farming.[25] The pottery was almost certainly used for cooking of both hunted and gathered materials. Ceramic crocks were probably used to carry and store water and raw foods. Perhaps they were also used as vessels for fermenting or preserving foods, including the production of alcoholic beverages. Modern humans think of alcohol primarily as a social lubricant, but it is in fact an excellent, calorie-rich way to preserve food that would otherwise spoil. Beer is, in many regards, a liquid loaf of bread.

As important as fire, cooking, and tools were to our transformations into modern humans, agriculture, once it took hold in societies around the world, tended to bring about huge changes in human culture. Just a few of these changes include a shift to permanent settlements and sedentary life-styles from a more migratory, nomadic existence; greater specialization by individuals, including the rise of full-time craft specialists, which then would have allowed for the elaboration and expansion of trades, arts, and sciences; increases in commerce and other aspects of economy; the formalization of political structures; an increase in wealth disparities between individuals; and changes in gender roles (which we will return to in chapter 7).

What, though, of the fishes in the parable of the loaves and fishes?

Stone tools, fire, and cooking have all been linked to changes in human anatomy and social structure; similarly, the consumption of fish, turtles, and other coastal foods may have been instrumental in the development of our large brains.[26] Coastal and riverine fishing is less dangerous and more accessible to people without sophisticated tools and communal hunting techniques than is the hunting of large terrestrial mammals.[27] Considerable evidence suggests that we migrated along coastlines for well over one hundred thousand years, which is consistent with aquatic fauna having been an important part of our diet.[28]

Chimps have been observed fishing for crabs in the Nimba mountains of Guinea. Extrapolating to early humans in similar habitats, we might infer that our early diet included a substantial amount of fish and other aquatic fauna.[29] The Beringians specializing in salmon on their route into the New World may have been improving on an ancient script. Fish eating may have been a critical piece of the puzzle of early human evolution.

A prehistoric world populated by millions of hunter-gatherers transformed within a period of ten thousand years into a world of one billion people consuming traditionally farmed food, and within the last two hundred years, we have transformed it further into a population of seven billion people surviving on intensive and unsustainable fossil-fueled farming, with which only a small percentage of the population has any direct contact. Those who have at least some relationship with the origin of their food—whether they grow it themselves, pick berries on occasion, or strike up conversations with the growers at their local farmers market—are more likely to value its complexity, the value of terroir, and the constant sharing between culinary traditions. People with some understanding of their food's origin and history are also less likely to assume that an energy shake is a complete replacement for food.

Harvest Feast

At the end of one of Heather's field seasons in northeastern Madagascar, where she was studying the sex lives of poison frogs, she was invited by the patriarch of a local family to photograph a *retournement*, a turning of the bones ceremony. During the annual ceremony, the bones of a select few ancestors are disinterred after the harvest. Those who have died recently, who were in body boxes, are rewrapped and put back in smaller bone boxes. Those already in bone boxes are given new shrouds. While the dead are out and about, though, the living take the opportunity to speak to them, to tell them of the major events of the last year—the size of the harvest and the frequency of storms, births, and marriages. Presumably, the dead are already aware of who has joined them on the other side.

On the day that Heather witnessed the event, the ancestors were already tucked out of view, but the ceremony would go on for nearly another twenty-four hours. It began with shots of *toaka gasy*, the local rotgut. This was followed by the ritual sacrifice of a zebu, the large-horned cattle that are common on the haut plateau of Madagascar, but are rare and precious in the wet lowland forests where we were. The slaughter itself was a quiet affair, watched by adults and children, after which the organ meats were laid out on banana leaves in the midday sun. A man and woman were assigned to "keep the spirits away" until the feast commenced in the early evening. To Heather's eyes, granted a relatively uninformed observer, the guards mostly seemed adept at keeping the chickens away.

An elder stood up and addressed the villagers—ancestors and living alike. He spoke in Malagasy, which neither Heather nor her field assistant could understand, but his effect on the audience was clear. His tone moved easily between sober respect and lighthearted memories, and his jokes were well received. He was clearly beloved. Sometime in the future, one of the living would be addressing him as one of the ancestors, in much the same way.

After the ancestors were addressed, the celebration became more raucous. In the hours before the nightlong feast would begin, there was music, dancing, and more *toaka gasy*. The women of the village danced in a long line, shaking their hips and singing, occasionally drawing in a man to their throng. The feasting that night, especially the meat of the zebu, would be remembered for a long time to come.

Madagascar is a land of feasts. Madagascar is also desperately poor. Everyday meals tend to consist of rice, ranonapango (burnt rice water, which is colloquially considered the national drink), and little else. One of the common greetings on the street, which even we, as obvious *vazaha*— white-skinned foreigners—sometimes received from strangers in our time on that giant red island, is "How many bowls of rice have you eaten today?" A high number is indicative of relative wealth, of being at least a bit removed from starvation.

Why do the Malagasy keep feasting when they are, as a country, starving? It's yet another paradox, which is, we argue, a kind of treasure map. When you see a paradox, keep digging.

Nature is not wasteful. So when you think you see waste—in the feasts of the Malagasy, in the massive temples of the Maya, in squirrels burying more nuts in the fall than they ever seem to dig up by spring—consider that you are likely looking with the wrong lens. There is likely a long-term strategy that is not visible with the normal tools.

Carrying capacity—again, the maximum number of individuals that can be supported in a given environment, at a given time—looks stable when you are zoomed out, looking across generations, or eons. Zoom in, though, and the perturbations in carrying capacity can be extreme, a parameter that oscillates wildly the closer you are in space and time. For agriculturalists, this looks like boom years and bust years: For every year that harvests overperform median expectations, there is another year that falls below the median. If birth rate were to track the noisy annual fluctuations in harvest, in half of all years there would not be enough to go around. Such years are naturally ones of conflict and division; long term, this is the death knell for a lineage. The solution involves productively spending excess resource so that it does not get converted into more babies, which will embody an unmeetable demand. Feasts are one such manifestation. By investing in community cohesion, rather than new mouths, a population can avoid the regular and predictable calamity created by the variance in harvests.

Tempering against boom and bust, a "fourth frontier" strategy of the sort we will return to in the final chapter of the book, has been a longstanding aspect of the human relationship with food.

The Corrective Lens

We might call this section *The New Kosher*. Most ancient dietary laws are now out-of-date, but that does not mean that we couldn't use some rules around how, what, and when to eat.

- **Shop the edges of the supermarket.**[30] Better yet, buy your food with as few middlemen as possible, as at a farmers market. Nearly everything from the middle of the supermarket has more sugar, more salt, more umami—generally by means that are not vetted, at least not in the long term. Chewing on sugarcane is to eating refined sugar as chewing on coca leaves is to snorting cocaine. Highly refined foods (aka "highly processed foods") are another example of hyper-novelty, as is plastic, so try to avoid food packaged in plastic, and especially avoid allowing hot plastic to touch your food.
- **Avoid GMOs.** GMOs are neither inherently dangerous nor inherently safe. They are, however, different from the artificial selection that farmers have been engaging in for thousands of years. When farmers choose plants or animals to breed, promoting some traits and down-regulating others, they are playing within the landscape that selection has already been acting on. In contrast, when scientists insert genes or other genetic material into organisms that have no recent history with those genes, they are creating an entirely new playing field. Sometimes they will be lucky, and the result will be useful and kind to humans. Sometimes they will not be lucky.

Chimerical life-forms that have been created by humans using hyper-novel techniques are not inherently safe; anyone telling you otherwise is either mistaken or lying to you.

- **Respect your food aversions and cravings**, especially after exercise, after illness, or while pregnant (so long as these cravings reflect real food and don't pose specific risks).[31]
- **Expose children to a diverse range of whole foods**, especially ones that connect them to your culinary and ethnic background. Eat the same food that you put in front of them, and show obvious enjoyment of it. Keep seasonal produce on your counter, and let the children eat any fruit that they find there, encouraging them to develop their own preferences while they also learn how and when to explore a variety of whole foods.
- **Consider your ethnicity and look to its culinary tradition for a guide to diet.** If you are Italian, look to Italian cuisine for clues as to how you should eat. If you are Japanese, look to Japanese cuisine. In particular, look to the culinary traditions of home cooking, as the foods represented in restaurants, while often delicious, often represent only a sliver of a culinary tradition's full panoply of options.
- **Do not reduce food to its component parts**—such as carbs and fiber, fish oil and folic acid. Instead, think of food as the species from which it came, the cultures that first used it, the myriad ways it is now prepared and eaten across the world.
- **Make food less ubiquitous in your own world.** For most of history, human societies have tempered against boom and bust with ritual feast and long periods of frugality. But recently, agriculture has led to an increase in the capacity to hold food in reserve, to save for a rainy day—or, more likely, to save for an extended drought, or a harvest failure. While our modern brains want to consume as much as possible, our ancient bodies want to store up for later. When calories were scarce and their availability unpredictable, this metabolic tendency made good sense. When a hunter-gatherer finds honey that he can separate from its bees, he and his friends will likely gorge on it, for there is no knowing when the next burst of sugary goodness will come their way again.[32] But since food resources are no longer scarce, gorging is not an effective

strategy, because the scarcity never comes. Instead, we get more opportunities to gorge. We have to willfully override our evolutionary impulses in order not to suffer the hyper-novelty that the twenty-four-hour grocery store provides. Putting yourself on a schedule, as intermittent fasting recommends, of not eating for regular periods of time seems to be a healthful corrective.

- **Do not forget that food is social lubrication for humans.** Eating alone in your car after visiting the drive-through is a novel situation, and it's not helping us connect with our food, our bodies and their needs, or one another.

Chapter 6

Sleep

SLEEP AT FIRST SEEMS LIKE A MYSTERY.

Were aliens to visit Earth, you might imagine that they would be confused by the fact that we go into a coma-like state on a daily basis, and hallucinate mad stories and strange characters with whom we interact while paralyzed. Except that they probably wouldn't be surprised at all, because any aliens capable of getting to Earth under their own power would almost certainly sleep, and dream, as well.

It's less certain, but still likely, that aliens who had managed to get to Earth would have gone through a phase, as we are now, in which their contemporary habits were out of sync with their ancient brains and bodies, causing widespread sleep disruption. Before mastering interstellar travel—before being able to do any of their best work—they would have needed to solve their sleep problems. Later in this chapter we will discuss some ways in which modern humans might solve ours.

For every animal about which scientists have asked, Do you sleep? the answer has come back in the affirmative,[1] which leads us to the question of *why*.

Sleep almost certainly came to be part of our lives as the result of a simple trade-off: it is impossible to build an eye that is optimized for both day and night. You could have two sets of eyes, but it would be impossible to build a visual cortex optimized for both without vastly increasing the brain's size and energy requirements. That creates a predicament. The predicament is: Should you be a creature with a very compromised eye, one that is not especially attuned to either day or night? Or should you specialize in one condition and sacrifice the other? All solutions exist. Day specialists are diurnal; night specialists, nocturnal; and those who specialize in the in-between times, crepuscular. There are diurnal dik-diks, nocturnal nightingales, and crepuscular capybaras. All solutions come with trade-offs.

All else being equal, if you have eyes, night is more difficult than day. Day provides an astronomical freebie. The sun broadcasts an enormous number of photons, which bounce off surfaces and accidentally reveal, to those with photoreceptors, where everything is. This is a huge gift. (Of course, we can also say with absolute certainty that being nocturnal also has its benefits, among them the lack of diurnal competitors. In any event, whether you are diurnal, nocturnal, or crepuscular, you will sleep for some portion of the Earth's rotation.)

We come from a many-million-year lineage of diurnal creatures. There are no nocturnal apes, and less than a handful of nocturnal monkeys, so our diurnality likely extends back at least to the most recent common ancestor of all simians (monkeys, *sensu lato*). So not only are there advantages to being a diurnal creature, because of the freebie that is light from the sun, but we also have a long history of diurnality in our evolutionary history. This raises the question, though, of what to do at night.

Sleep saves energy. If your eyes aren't adapted to the night, trying to be ecologically productive then is likely to be both wasteful (as you'll miss the things you need to see) and dangerous (as night-specialized hunters will likely be better at finding you than you are at avoiding them). Given the hazard that starvation poses to every animal, if one is not going to be productive, figuring out how to be dormant is a priority. Not wasting energy is, in some regards, as valuable as finding energy. The question then becomes, How dormant do you need to be?

For humans, in particular, it would be a shame to take the marvelous computer that we have riding around on our shoulders and totally sideline it during the night, just because our eyes are out of their depth. Because even when we can't literally see what is going on in the world, we can consider what we've already seen. In response, selection has borrowed the incredible computing power that exists in our visual apparatus and repurposed it for a kind of moviemaking. We are physically dormant at night, but not mentally so.

In sleep, we predict and imagine what we might see in the future, and do a bit of scenario building around the possibilities. So that next time, we know what to say, how to feel. So that next time, we are prepared.

We can predict that intelligent aliens would recognize sleep right off the bat, because although the Earth is special in many ways, day and night are shared features of all planetary bodies on which life is likely to evolve.[2] If aliens are complex, smart, and social enough to have visited us, they likely come from a planet with a similar predicament, one with light-filled days and dark nights, one in which physical dormancy for part of the day has made historical sense, but during that time, the mental apparatus has become highly active.

Broadly speaking, sleep as we know it can be broken into two types: rapid eye movement (REM) sleep, in which the eyes flit rapidly around and the muscles of the limbs and trunk go limp, producing paralysis; and non-REM sleep, the deepest form of which is slow-wave sleep, during which brain wave activity slows and synchronizes.[3] All animals appear to sleep, but only mammals and birds have REM sleep—although the first stirrings of REM sleep have been observed in an Australian lizard.[4] Slow-wave sleep is thus more ancient than REM, and in species that experience both types, it occurs earlier in the night as well.

During slow-wave sleep, our brains fix memories in place—as do the brains of great apes, including chimpanzees.[5] Our brains also prune out old and redundant information during slow-wave sleep and gain mastery over skills that we learn while awake—typing, skiing, calculus. Hence the adage to "sleep on it." REM sleep, the newcomer in evolutionary time, provides us our dreams. During REM sleep, we engage in emotional regulation, reflect on what has happened, look forward to what might be possible, and imagine both possible pasts and futures. REM is a creative state; it is sleep's explorer mode.

REM can be chaotic and disorganized in its creations, and it might be said that slow-wave sleep acts as a corrective to some of REM sleep's products. Once selection discovers that there are useful ways to utilize a mind during bodily dormancy, it discovers all kinds of utility, and sooner or later, individuals become dependent on their ability to access this state. Our bodies and brains, linguistic and emotional lives, social and behavioral repertoire are all dependent on sleep.[6]

Slow-wave sleep is ancient, going back at least to the origin of animals, and is necessary for all kinds of repair. Thus, the benefits of sleep are far more ancient than those of dreams, but our dream state is so beneficial to us in scenario building that it has far outstripped, in the positive direction, the risks of sleep. The upsides of sleep are greater than the downside of spending a third of every day physically dormant.

Dreams and Hallucinations

In the darkest, quietest moment of some night long ago, hours after we had both been asleep, Heather sat up, looked at Bret, and said, "Do you seriously intend to leave these car parts in the bed?"

Bret's answer—"I think so, yes"—did not ease tensions any. Furthermore, the fact that there were, of course, no car parts anywhere near the bed would not be admitted as evidence in this discussion.

It wasn't the first time that Heather had said something while deeply asleep that was impossible to engage using the normal rules. When Bret responded to Heather's sleep talk, the tone generally decayed quickly. There was, it seemed, no reasoning with either of us. Without having any consciousness of these episodes, Heather somehow knew—later, when awake and being told about them—what Bret was supposed to do.

"Don't engage me. Let me have both sides of the conversation, and it will be over soon enough."

Seeing things that aren't there. Hearing sounds that were never made. Believing things that are not true—yet being certain of them. Being unable to control one's movements. Having conversations with people who don't exist.

As it turns out, a list of symptoms of a person with schizophrenia has a suspicious overlap with a person asleep and dreaming: all of us enter this state every night, even though not everyone reaches out through that state and talks in their sleep. We do not regularly draw this parallel, because our dream state usually comes with paralysis and amnesia. Any confrontations with reality are blissfully hidden from us by the time we get to our morning coffee.

How surprising, then, that organisms that do not appear to have had our best interests in mind—such as *Psilocybe* mushrooms and peyote cactus—seem to have accessed these very same tendencies.

To explain this, we need to take a step back. Organisms—including us and other animals, plus plants and fungi—do not generally want to be eaten. Fruit, nectar, and milk are, as we discussed in the previous chapter, exceptions to this rule, but in general, organisms put a lot of work into discouraging the consumption of their body parts. Structural barriers are one method—cactus spines, porcupine quills, turtle shells. Another is poison, but it is often too crude to be maximally effective. If a deer dies after eating foxglove, the deer will be replaced by another deer that knows nothing of the plant's poison. On the other hand, if a deer expands its dietary repertoire to include *Psilocybe* mushrooms, and spends the rest of the day having a temporary psychotic break, it may well look elsewhere for its next meal, having been educated, and perhaps terrorized, rather than killed.

Secondary compound is a loosely defined botanical term for a substance that is not functional within the organism that produces it. Rather, it is intended to interact with pathways in other creatures, often in a hostile way. The irritants in poison ivy are an obvious deterrent to herbivores eating those leaves. Similarly, potatoes and the other nightshades contain endogenous pesticides, a class of compounds known as glycoalkaloids that are highly toxic to humans. In contrast to those pure poisons and irritants, consider these secondary compounds: Capsaicin, the molecule that creates the burning sensation when we eat chili peppers, generally dissuades mammals from eating seeds that are intended for birds, which do not have the receptors to sense the "heat." And caffeine, which disincentivizes herbivores from eating caffeinated seeds at high concentrations, may also be a kind of pharmacological social engineering on the part of the plants. When bees are given sugar rewards that contain caffeine, their spatial memory improves threefold; the caffeinated nectar of both citrus and coffee flowers may well be priming their pollinators, the bees, to remember them and to come back for more.[7]

From *Psilocybe* mushrooms and ergot fungi, to peyote cactus and the botanical brew in ayahuasca, to salvia and Sonoran Desert toads, there are fungi, plants, and animals that have produced secondary compounds that interact with our physiology in ways that mirror dream states. Call them hallucinogens, or psychedelics, or entheogens—their effect on us can be narrative and elucidating.

We live our days, connected by dreams. Lest we wake up each morning imagining that we were brand-new beings, our dreams give us context, and allow us to grow between days.

We are conscious during the day, and unconscious in the first part of the night, during non-REM sleep. Once REM picks up in the second half of each night, our consciousness is borrowed, our bodies taken safely off-line, paralyzed, and our conscious minds create fictions strange, hypothetical, and extravagant. Sometimes they are even true.

A vast array of cultures have some tradition in which hallucinogenic states are intentionally triggered in some or all of their members. Humans being human, it is not surprising to find that many cultures have borrowed secondary compounds that trigger terrifying waking dreams, and have turned what might have been a bad trip into an important tool for human consciousness expansion. More on this in chapter 12.

Just as many cultures have appropriated the hallucinogenic secondary compounds of plants and fungus to expand the consciousness of their members, many also have sleep rituals—from the simple to the elaborate—by which individuals prepare for their nightly slumbers. Even some of our closest relatives engage in rituals in advance of sleep.

Nightfall in the Jungle

Night was falling over Tikal. Now a grand ruin with jungle encroaching on all sides, Tikal was once a center of commerce, politics, and agriculture for the Maya.

Dusk in the rain forest is a time of transition—the diurnal are slowing down for the day, the nocturnal are waking up, and the crepuscular are skulking about looking for opportunity. This is their moment. The sounds of the day—birdsong and the endless throbbing drone of cicadas—disappear as choruses of frogs begin to thrum, and myriad spiders become visible by the red shine of their eyes. Because no animal is equally active in the day as in the night, at dusk, as at dawn, the cast changes.

It was the early 1990s, and we were in the middle of a long backpacking trip through Central America, near the base of one of Tikal's temples, as the shadows grew long. Dusk fell fast, as it does in the tropics. This was long enough ago that camping right next to the temples was allowed. We found a place to pitch our tent as the light failed, and talked about our day.

Then the spider monkeys arrived. High overhead in the canopy of the rain forest, they, too, were bedding down for the night. They, too, talked to each other. Linguists and others may object to this characterization though, as it's not quite right. Spider monkeys are not known to have syntax, or much vocabulary, or many of the other expectations of linguistic exchange. They were surely communicating with one another though, chattering. We stood on the ground, taking a break from establishing our own temporary home for the night, watching our glorious primate relatives go through their nighttime rituals.

The rituals of spider monkeys are protective, hiding or otherwise immunizing them from nocturnal predators. They may have sentries, someone who sleeps near the outside of the group, watching for intruders.

For our recent ancestors, protective nocturnal behavior had a lot to do with fire. Earlier humans gathered around fires and engaged in a highly unusual activity: We talked. We exchanged information about the day, our views about the future; we recounted stories passed down from the ancients. We sang. Sometimes we danced. Then we went to sleep.

Those ancestors of ours, like the spider monkeys, were a community moving from dinner to sleep. They did not have the sleep difficulties that many 21st-century humans do. They fell asleep easily, slept well, and woke refreshed.

Novelty and Sleep Disruption

So far as we know, monkeys do not tap into their dream state with the use of entheogens. This uniquely human kludge can be viewed as humans using novelty to potentially adaptive ends. There are, of course, many ways in which our relationship with sleep is suffering as a result of novelty. Electric lights are top of the list, a list that also includes air travel, noise pollution, and a twenty-four-hour economy that has many people working night shifts.

Tucked deep in our brains is a region called the suprachiasmatic nucleus, which acts as our biological clock. It keeps track of what time of day it is—not in the sense of "5:00 p.m.," but in the sense of where we are with respect to photoperiod, the length of time in a twenty-four-hour period during which there is light—because that was, until very recently, the only important parameter. In London, 4:00 p.m. is called daytime in both December and June, even though in June the sun is still high in the sky at 4:00 p.m., whereas in December the sun has already set. Until recently, darkness mattered far more than one's position in a twenty-four-hour day. So we humans have done the expedient thing. We've used ingenuity to extend our productive period, by inventing artificial light. The benefits of this are obvious, but the hazards aren't.

Before electric lights were invented, humans never experienced light after sunset of the intensity or duration that we are now commonly exposed to in our indoor spaces.

Daylight is intense, even on overcast days, but our brains do a fantastic job of obscuring from us just how bright or dim our surroundings are. Anyone old enough, or retro enough, to have shot film (rather than doing digital photography) will remember being surprised by light-meter readings that were widely variant between settings that all seemed plenty bright. Our experiences in this were brought to the fore in the rain forests of Latin America and Madagascar in the 1990s. The understory of a lowland tropical rain forest is a dense, lush tangle of vines and shrubs, punctuated by massive, buttressed tree trunks and the sounds of insects. Rain forests don't seem like particularly dark places, until you reach an edge and walk out into a logged pasture, or onto a road, and find yourself blinded by the sun. Light meters don't lie, and they reveal that the percentage of light on the rain forest floor is a tiny fraction of that at the top of the canopy—just 1 percent by many measures. Yet our eyes adjust, and we can see just fine in such conditions.

What this tells us is that we are ill-equipped to know when the level of light we are being exposed to is outside of normal. While daylight is bright and tends toward the blue end of the visible spectrum, and moonlight and firelight are dim and tend toward the red end of the spectrum, indoor lighting is typically brighter than either moonlight or firelight, and far bluer than, but not as bright as, daylight. This has the potential to interfere with circadian rhythms and hormonal cycles, and therefore to cause sleep disruptions. It has now been demonstrated that evening light in the intermediate intensity range causes circadian disruption (and it is also true that differences between individuals in susceptibility to such disruption are high, making it difficult to extrapolate from anecdote to whole populations).[8]

By contrast, humans have such a long history with fire that our pineal glands are well equipped to encounter red, fire-spectrum light well past sundown, without negative consequences for sleep. Being able to turn on blue, day-spectrum light at any moment, however, is a brand-new phenomenon, one for which we are less well adapted.

Empirical science has begun to reveal just this: blue-spectrum light is unhealthy at night.[9] Recently, the market has responded with an abundance of red filters and software that change the spectrum emitted by screens at night. We could have figured this out far earlier if we had applied the proper precaution to the invention of lightning in a bottle. It is obvious that the ability to generate light at the end of a wire has transformative potential. The chances that it would be harmless approach zero.

And now we are making that same error again. The transition from tungsten-based light bulbs through fluorescents and into LED has again pushed us in the direction of colder, bluer light, characteristic of daytime. Worse than that, for many WEIRD 21st-century people, little blue LEDs now shine or blink in every room of our homes (unless they are covered or otherwise obscured—which we recommend). Our brains evolved being able to intuit the time of day by the spectrum of light coming into our eyes. Now we have midday blue flashing at us at all hours; no wonder sleep has become elusive for many.

If we really understood the costs and benefits, we as a civilization would tightly regulate light spectrums so as to leave our sleep and wake cycles intact. Many suffer from insomnia at home but watch it vanish on a camping trip, as the cycle guided by the light of the sun and moon returns us to a more ancestral state. We hypothesize that eradicating daylight-spectrum light from the night might have a curative impact on some people who are experiencing debilitating psychological disorders, those who suffer from daytime delusions, paranoia, hallucinations—those who are, it might be said, having their dream state intrude on their waking day.

As if that is not enough to concern us, we are hardly the only organisms that have shown a deadly sensitivity to electric light.

Everyone has watched moths pointlessly captivated by a light bulb. They do this because they are wired for a nontechnological world. They may be navigating based on the angle that they are flying relative to the moon, which until recently would have been the only large, bright object in the sky; or they may be trying to escape the light and failing abysmally.[10] Whatever the reason, when we place other bright objects in their world, their program has a devastating effect—they fly, holding those objects at a fixed angle, circling to the point of exhaustion.

Sleep-wake cycles in wildlife are also altered where light pollution occurs, and many biological rhythms and behaviors desynchronize when exposed to light at the "wrong time" of day.[11] Many organisms, especially far from the equator, use photoperiod as a clock. It is used to schedule such things as germination and bud formation in plants; and mating season, molting, and embryonic development in animals.[12] And animals as distantly related as crows, eels, and butterflies have difficulty migrating when artificial light is present.[13]

Electric lights dramatically change the when, where, and what (spectrum) of light historically available to us. Not only are we confusing our own brains and making ourselves sick with electric lights, but we are deeply confusing many other organisms as well.

These last four chapters have focused on the survival of our individual bodies in the realms of health and medicine, food and sleep—in a world wherein it's becoming increasingly difficult to do so. The evolutionary story of humans is not primarily one of individuals surviving, however; it's one of people coming together. Indeed, the extent to which we are interconnected may confound and challenge the modern imagination. In the next section, we move to things larger than the individual—sex and gender, parenthood and relationship.

The Corrective Lens

- **Allow celestial bodies to set your sleep-wake pattern.** Wake with the sun. Know the phases of the moon. Navigate sometimes by the light of the full moon. Navigate sometimes as dawn emerges, or as dusk falls, paying attention to the shifting of your senses as light becomes more, or less, available. Spend time outside, letting your body take cues from the light of the sun, rather than from the light switch on your wall or the screen you gaze into.
- **Get closer to the equator at some point during your winter.** Cross the equator and keep going for ever more light during the dark days in your home hemisphere. Especially if you are susceptible to seasonal depression, you probably live relatively far from the equator, such that winter brings months of darkness in the form of short days and a low sun angle. This advice, of course, comprises a novel opportunity (global travel) that is itself a response to the novelty of living indoors with electric light. When we were in grad school in Michigan, Heather had several good scientific reasons to conduct field research in Madagascar—which is in the Southern Hemisphere, off the east coast of Africa—between January and April. A lovely collateral benefit of this fieldwork was that she effectively cheated her photoperiod allotment, heading to southern summer when the northern winter was at its darkest.
- **Avoid caffeine less than eight hours before bedtime.** Children and adolescents are better off with no caffeine at all, because of its strongly sleep-disrupting effects and the nonreversible effects of

sleep deprivation on the developing brain.[14] Similarly, avoid pharmaceuticals as sleeping aids—we don't know what all they do, and
we do know that they are often disruptive to actual sleep.

- **Go to sleep early enough that you wake without artificial help**—with
  the sun beginning to come in your windows, for instance, rather
  than with an alarm that bursts into your consciousness and disrupts
  your dreams.

- **Develop a ritual in advance of sleep**, just as the spider monkeys at
  Tikal do. It might be as simple as turning lights down low as bedtime
  approaches, or more elaborate, but a regular series of behaviors
  can come to signal to your body that it is soon time to sleep.

- **Spend time outside every day**—sunlight calibrates your sleep-wake
  cycle far better than artificial light does.[15]

- **Keep your bedroom dark while you sleep.**[16] This includes removing, turning off, or covering all blue indicator lights from devices.

- **Use a red light, rather than a standard one, if you read before
  bed.** Whether you are or are not highly susceptible to circadian
  disruption from medium-intensity, blue-shifted light in the eve-
  nings, there is a good chance that someone in your home is.

- **Restrict outdoor blue-spectrum light at the societal level**, particularly lights that shine upward and outward at all hours of the night.
  Nighttime darkness is healthy; twenty-four-hour light is not, and is
  even implicated in higher rates of disease.[17] Furthermore, humans
  deserve a night sky, a sky full of possibilities—sometimes of clouds,
  often the moon, occasionally planets, nearly always stars and the
  Milky Way in which we live. Besides sleep, which we need, what
  else might we lose when we disappear our own night sky?

Chapter 7

Sex and Gender

IT'S 1991 IN MANAGUA. WE HAVE BEEN TRAVELING THROUGH CENTRAL America all summer—narrowly missing a full eclipse in southern Mexico after driving all night to see it; watching those monkeys prepare for sleep in the shadow of Tikal; snorkeling and sleeping in hammocks on a tiny cay that we had all to ourselves for three days off the Caribbean coast of Honduras. Now we are in Nicaragua, meandering through a large outdoor marketplace, wandering apart whenever a fruit Bret hasn't seen before catches his eye, or the smell of fresh baked goods catches Heather's attention. We are both comfortable being alone; neither of us sees coming what happens next.

Heather finds herself suddenly surrounded by a crowd of young men, young men who seem like all hands and arms, reaching yet not ever quite grabbing or groping. There are eight or ten of them, and they begin to move all in the same direction, pushing Heather along with them toward the edge of the market. Heather begins to shout, and the young men continue to move her with them, but Bret shows up fast. He yells at them, and they stop. Heather pushes herself out of the mob and stands away from them, breathing hard.

Then the young men line up, all of them, and one by one they apologize to Bret.

Heather is furious, as is Bret. We are also both surprised; as much as macho culture is real, we hadn't seen anything like this before. We have just gotten a view into some traditional gender norms, one in which women are viewed as the property of men. You apologize for trying to take someone's property; you don't apologize to the property.

Our work in evolutionary biology has since revealed to us that the roles of men and women have been distinct for most of human history, but this was the first time that either of us was face-to-face with some of their most unfortunate manifestations. Regressive behavior like this is familiar from history and literature both, and leads many people today to believe all traditional gender norms as regressive. This thinking, however, is a mistake.

When sex was scarce—or at least, less ubiquitous than it is now—men would move mountains for the right woman, with immense societal consequences. The Taj Mahal was built in memory of the Mughal emperor's favorite wife. Odysseus, returning after twenty years away in war and voyage, won back Penelope's love by passing a test of skill (and killing her other suitors). And, of course, there is Helen of Troy.

No group of young women would surround a young man in the same way that the men surrounded Heather in that Nicaraguan marketplace. If they did, considerable numbers of young men would appreciate it. Wars don't start over the love of a good man. Temples aren't generally built to impress husbands. None of this happens if the sexes of the players are reversed. What happened in that marketplace is offensive to modern mores because it reveals a belief in women as resources to be exchanged, rather than as whole entities with all attendant desires and aspirations. Modernity does not abide that; nor should it. But some traditional gender norms will be more tenacious than others.

Today, men and women work side by side in nearly all domains. Both sexes have broken through boundaries once thought impossible to break, to the benefit of individuals and society alike. Some of the population-level differences that have long been attributed to men and women turn out to be mutable—women should not be confined to healing or teaching professions, nor men to ones requiring brute strength or raw ambition.

Recognizing these things does not mean, however, that we are the same at the population level. For instance, "Men are taller than women" is a true statement about averages. An average difference does *not* imply that all members of population Y (men) are taller than all members of population Z (women). True statements about populations do not manifest in all individuals in those populations; believing otherwise is falling prey to the "fallacy of division" (first described by Aristotle). In populations where the overlap of a trait is significant, it can be difficult to parse population-level patterns from individual experience. If you, as an individual, do not fit a particular pattern, the discrepancy can feel like evidence that the pattern is false, but that feeling does not make it so.

In professions from medicine to sales to soldiering, men and women work together, but are we really doing the same thing? Female doctors are more likely to go into pediatrics; men are more likely to become surgeons.[1] In retail, men are more likely to sell cars, women are more likely to sell flowers.[2] And while retail jobs in the US were split nearly evenly between men and women in 2019, wholesale jobs skewed strongly male.[3] In tasks that require physical strength, men, on average, are simply stronger. An all-female force engaging in hand-to-hand combat would not beat an all-male one, and it would be beyond foolish to pretend otherwise.

We work side by side, and some of us imagine that because we are equal under the law, we are also the same. We are and should be equal under the law. But we are not the same—despite what some activists and politicians, journalists and academics would have us believe. There seems to be comfort, for some, in the idea of sameness, but it is a shallow comfort at best. What if the best surgeon in the world was a woman, but it was also true that, on average, most of the best surgeons were male? What if the top ten pediatricians were women? Neither scenario provides evidence of bias or sexism, although those are possible explanations for the observed patterns. In order to ensure that bias or sexism is not predictive of who does what work, we should remove as many barriers to success as possible. We should also not expect that men and women will make identical choices, or be driven to excel at identical things, or even, perhaps, be motivated by the same goals. To ignore our differences and demand uniformity is a different kind of sexism. Differences between the sexes are a reality, and while they can be cause for concern, they are also very often a strength, and we ignore them at our peril.

Sex: The Deep History

We have been sexual beings for at least five hundred million years, since long before we were human. That number may well be an underestimate, too—it is quite possible that we have been sexual since becoming eukaryotes, anywhere from between one and two billion years ago.[4] That is a very long time indeed. Without interruption, our sexually reproducing ancestors go back not millions, not tens of millions, but many hundreds of millions of years.

Sexual reproduction has always been a messy, costly operation. You have to find an appropriate mate. You have to convince the mate that you're a good bet. It may need to be the right time of year—the mating season—or else it's possible that your gonads have been reabsorbed to save weight and to use those resources for something else (many migratory birds do this—male song sparrows have essentially no testes while they're migrating long distances, and they (re)grow a pair upon landing in their mating grounds). If you do manage to find and convince another individual of the right species and breeding status to mate with you, you may have to tend to the developing egg or fetus. You may have responsibilities that extend for years—decades even—after you have a child that is the result of sexual reproduction.

Compare these to the biggest cost of all: In terms of genetic fitness, when you reproduce sexually, you are taking a 50 percent hit. If you cloned yourself, you would be 100 percent related to all of your offspring, spreading your genes with perfect accuracy. With sex, only half of your genes are in each of your kids.

Given all of these costs, why on earth did sexual reproduction evolve? And why did it stick?

While actively discussed among scientists still, the broad answer is that *asexual* reproduction is only a win for you and your offspring if the future looks exactly like the past.[5]

So long as conditions stay the same, if life worked out for you, it should work out for your clones.

But conditions don't stay the same, do they? Some change is predict-able, like seasonality, for instance. Most change isn't—can you predict when and how bad the next giant flood will be, or harvest failure? Mixing up your genotype with someone else's, possibly breaking up some bad genetic com-binations that had been riding around in you, perhaps discovering new good combinations, and giving your offspring a chance at being a better fit in a landscape that has not yet occurred—these are the benefits of sexual reproduction.

In alligators, the temperature of an egg as it develops determines the sex of its inhabitant: low temperatures create females, high temps make males. The same is true for tortoises, but the outcomes are reversed—cool eggs produce males, warm ones produce females. And in crocodiles and snap-ping turtles, intermediate temperatures produce males, while either ex-treme produces females.

In mammals and birds and a smattering of other animals, by comparison, sex is determined chromosomally. In all but a very few species of mam-mals,[6] females are XX, males are XY.[7] Unlike in some organisms with en-vironmental sex determination—famously, clown fish—no mammals (or birds) have ever been known to change sex. When human sperm fertilizes an egg, all the genes in the zygote that aren't on the Y chromosome discover the sex of the individual they will become. By ratio, our genome is over-whelmingly sexless, the presence or absence of a single chromosome making the determination. But the fact of our overwhelmingly sexless genome does not imply that distinctions between the sexes are either small or arbitrary.

Our chromosomes start us down a path toward femaleness and male-ness. One of the genes on the Y chromosome, for instance, is *SRY*, which, when initiated, controls the regulation of a whole series of masculinizing actions, including the formation of what will become the testis, in which sperm is made. Hormonal cascades further masculinize a body (with tes-tosterone and other androgens) or feminize it (with estrogens and pro-gesterone). Even when the quantity of gonadal hormones is controlled for, though, sex chromosomes *themselves* affect such varied differences between males and females as pain perception and response, the anatomy of individual neurons, and the size of various brain regions, including parts of the cerebral cortex and corpus callosum.[8]

All of that is true. All of these explanations are true, mechanistic, proximate explanations of how individuals become female or male in mammals. Yet these explanations can get nowhere in explaining why male and female exist at all. For that, we need ultimate-level explanations. Evolutionary explanations. Explanations that begin with the question, *Why?*

Why are there two sexes in nearly all of the sexually reproducing organisms on the planet, and not three, or eight, or seventy-nine? Fungus does things rather differently, but among plants and animals, there are only ever two types of gametes (reproductive cells).

In order to reproduce sexually, you need two things: DNA from multiple individuals, and a cell. The machinery of the cell—for example, mitochondria and ribosomes—is big and bulky, at least compared to DNA, but it's necessary for life. So if you are going to reproduce sexually, at least one partner has to contribute that cellular machinery, known as the cytoplasm. That cell—which we call the egg—is therefore big (for a cell). Trade-offs being what they are, that big cell is also largely sessile—it doesn't move.

The next problem in sexual reproduction is how the gametes will find each other. Since some gametes are sessile, others need to be zippy, which is facilitated by being stripped of most of the cellular machinery that would be required to make a zygote. Those gametes are called sperm in animals, pollen in plants. They move around their environment, "looking for" eggs. Intermediate gametes—ones with some cytoplasm and some ability to move—would be worse at both things: they would have insufficient cytoplasm to create a zygote on their own (and upon meeting a gamete that also had cytoplasm, there would be disagreement about whose to use to make a whole new life), and they would be slow to find other gametes to hook up with. Anisogamy, the evolution of two different (*aniso*) gamete sizes (*gamy*), therefore occurs because of the poor performance of intermediate gametes.

Fast-forward many hundreds of million years, and differences between the sexes abound. Humans are sexually dimorphic across many domains, extending far beyond reproduction. Men and women have different disease risks, etiology, and progression, for conditions from Alzheimer's[9] to migraine,[10] from drug addiction[11] to Parkinson's disease.[12] Our brains are structured differently.[13] We tend to have different personality traits by sex, and they are mediated by our environment: personality differences are greater in countries that have abundant food and a low prevalence of pathogens.[14] In general, women are more altruistic, trusting, and compliant as well as more prone to depression than men.[15] Men are more likely to be diagnosed with ADHD,[16] women are more likely to have anxiety disorders.[17] Finally, on average, men prefer working with things, and women prefer working with people.[18]

It is no accident that, in every human culture known, there is language that distinguishes male from female.[19] It's a human universal.

Sex Changes and Sex Roles

Sometimes, conditions are so dire that normally sexually reproducing individuals become asexual in order to reproduce. Among vertebrates, this has been observed in a few snakes and a hammerhead shark.[20] Female Komodo dragons—which, strictly speaking, are not actually dragons, but very large lizards that live on rather small islands in eastern Indonesia—have also been known to produce viable eggs despite having had no exposure to any other Komodo dragons.[21] This, presumably, is an adaptation, a last-ditch response when you've shown up alone on an island and have none of your kind with you. It's not optimal, but it's better than nothing.

Similarly, sometimes conditions are such that it is evolutionarily appropriate to change sex. In a few species of plants, many species of insects, and several clades of reef fish, sequential hermaphroditism is common, in which an individual begins as one sex and switches to the other at some point in their life. Flame wrasses,[22] for instance, may begin life as females but transition as large adults into particularly colorful males that attract most of the sexual attention from females. Among the tetrapods, however—those vertebrates that came onto land back in the Devonian Period—there is only a handful of species that are known to have switched sex,[23] and only one that does so with regularity, the African reed frog.[24]

Among sequential hermaphrodites like flame wrasses, after a female changes into a male, he has changed not only his sex—which gamete he produces, once eggs, now sperm—but also his "sex role," which is the behavioral expression of his (new) sex. In humans, we call this gender, or sometimes, gender expression.

In moose, the sex role (gender expression) of males includes engaging in showy combat, and injury is not uncommon. In golden-collared manakins—neotropical birds—the sex role of males includes clearing an area on the forest floor and dancing on it. In great bowerbirds, the sex role of males includes building elaborate bowers—temples, if you will—which not only contain an array of carefully chosen items but even use forced perspective, as artists do, which makes the bowers seem bigger than they are from the direction that females approach.[25] In all of these species, the sex role of females includes choosing from among the males—the combatants, the dancers, the temple builders—and raising the children, be they moose embryos and calves, or bird eggs and chicks.

The usual rules of sex roles, then, are ones of male display and female choosiness. This stems from that long-ago difference in investment between the sexes—the large resource-rich egg and the small streamlined sperm. Furthermore, in those species in which parental care is necessary for the offspring to survive—which means all of mammals and birds, and a good proportion of reptiles, amphibians, fish, and insects as well—males tend to put more effort into what happens before sex, and females put more effort into what happens afterward.[26]

To put it in strictly evolutionary terms, in the vast majority of species, females are the limiting sex. Because females invest more in offspring—from eggs being larger than sperm through parental care typically (though not always) falling to females more than males—males must compete for access to females, and females get to choose among their suitors. Males thus tend to be larger (think elephant seals) or more aggressive (e.g., woolly monkeys); or gaudier (peacocks), louder (nearly all frogs), or more melodious (mockingbirds) than the females of the species.

There are a few rare species out there that we call "sex-role reversed," in which the usual rules of male display and female choosiness are flipped.

Sex-role reversed species have also flipped which sex invests the most when; in these cases, the male is the limiting sex. Several polyandrous waterbirds do this, including Northern jacanas. Among them, we find dominant females defending large territories, inside of which a female's many male mates build nests, incubate eggs, and tend to the young. While humans don't start wars over the love of a good man, and generally don't build temples to impress husbands, in sex-role reversed species of birds, these things might just happen.

Still, sex-role reversal—what we might call gender switching in humans—is not the same as changing sex. In mammals and birds, with our genetic sex determination, there is no sex change possible—no pigeon or parrot, no horse or human, has ever changed what sex they actually are. Behavior, though—call it sex role, call it gender—*that* is highly labile (open to change). We humans are the *most* labile, behaviorally, of all the animals. So it shouldn't surprise us too much that many of us are abandoning some old gender norms—behaviors that have, in the past, been tightly coupled with our sex—and configuring new ones.

What it would be foolish to do—and what many WEIRD people in the 21st century are doing—is to pretend that sex equals gender, or that gender has no relationship to sex, or that either sex or gender is not wholly evolutionary. Remember the Omega principle, which tells us that adaptive elements of our software (e.g., gender) are no more independent of our hardware (e.g., sex) than the diameter of a circle is independent of that circle's circumference. Gender is more fluid than sex, and has many more manifestations, but "acting feminine" (gender) is not the same as "being female" (sex).

If you want, be a woman who gets into bar fights or a man who wears makeup, but don't imagine that getting into bar fights makes you a man, or that wearing makeup makes you a woman. Bar fights and makeup are signals to the outside world, proxies. Proxies are not the thing itself, and these particular proxies are also outdated and regressive. Some of gender, however, is neither outdated nor regressive: women are, on average, more likely to nest and nurture, and men are more likely, on average, to defend and explore. Observing this does not mean that men don't nurture, or that women don't explore, but these population-level differences have evolved because of the underlying differences between the sexes. Pretending otherwise puts us all at risk—ask people to believe things that are patently untrue and they will be ever less likely to form a coherent worldview, one based in observation and reality, rather than fantasy. Men will never ovulate, gestate, lactate, menstruate, or go through menopause. Women who identify as men might, but that is different.

Sexual Selection in Humans

The moose's antlers, and the fights that they have, and the fact of female choice—these are all sexually selected characters. The egg incubation behavior of male American jacanas—this, too, is sexually selected.[27] The radically different size of male and female elephant seals, the fact that only male frogs call, and the fact that from peacocks to quetzals to mallards, males have far showier plumage than females—all sexual selection. In the next chapter, we'll consider how mating system—monogamy versus polygyny, mostly—affects sexually selected characters, but for now, let us consider a few ways that men and women reveal the effects of sexual selection.

Human girls develop breasts at puberty, and they persist throughout the woman's life. Breasts are, of course, functional as mammary glands with which to feed babies. In no other species of primate do breasts persist when there are no babies around to benefit. Human breasts are sexually selected, and they are doing more than feeding babies. They are also advertisements to men—just as a lyrebird's song and a rutting boar's smell[28] and a red-capped manakin's dance are advertisements to the females of those species.

The concealment of ovulation in humans is also sexually selected. While nearly all mammals advertise fertility by physiological means, humans do not—or at least, we do so far less than other species. We have also become sexually receptive throughout the year, rather than just seasonally. Concealment of ovulation serves some reproductive ends, but it also encourages something that humans do a lot of: we engage in nonreproductive sex—sex for pleasure, sex for bonding.

What else is sexually selected in humans? Flowers on her birthday. Neckties. Fast cars. Makeup and heels and jewelry.[29] Indeed, the physical

ornamentation of women—including not just makeup and heels and jewelry, but also breasts that stay enlarged throughout our reproductive
cycles—is an indicator of partial sex-role reversal in humans. What does
that mean? Whereas most animal species exhibit male-male competition
and female choice of mates, a species with partial sex-role reversal, such
as we have, will also show competition between females and male choice
of mates. This can manifest as anything from women advertising to attract
the attention of men to outright brawls between women. Men, not coincidentally, will also be more likely to be in a position to choose their partners.

Division of Labor

In many modern households, women clean the floors and men take out the
trash. In some households, those roles might be switched, and it might be
true that both members of a pair-bond spend equal time doing domestic
work, but it's fairly rare for both partners to do every domestic task in equal
measure. This is division of labor.

From many angles, the division of labor makes sense. It has even been
argued that division of labor by the sexes is what made us human.[30] Even
if we don't accept that conclusion, we can agree that it's efficient, and generally a good use of everyone's time—saving time by dividing up the work
leaves more time for things we might want to do more of, like play or sex.
The division of labor can and has created rigid roles, however, many of
which are outdated in the 21st century. It is useful to understand some of
where those roles came from in order to determine which ones are unlikely
to change, and which ones might.

From the earliest inequalities in investment in gametes, females and
males have engaged each other, and the world, differently. Among hunter-
gatherers, men have been far more likely to be hunters of large game,
women more likely to be gatherers of plant foods and smaller animals.
Hunter-gatherer women likely spent most of their adult, premenopausal
lives pregnant or breastfeeding infants and toddlers. When breast milk is
all or most of a child's diet, the mother is effectively on birth control, as
she experiences physiologically induced amenorrhea—she cannot get

pregnant when breastfeeding at frequent intervals. This keeps birth intervals relatively long, and the birth rate fairly low.

Jump forward to the human transformation of landscapes with agriculture, and gender roles became even more constrained. Being tied to a particular piece of land, we were now more sedentary and had ample grain stores with which to supplement our and our children's diets at any time. Agriculturalist women thus experienced a decrease in the birth interval—babies came at a faster rate—and so the birth rate climbed.[31] This increase in fertility tied women to hearth and home, and we saw a concomitant decrease in women's roles in economic, religious, and other culturally important realms.

Men and women exhibit so many differences, it would be impossible to catalog them all here. Before we mention just a few more, another reminder about populations is in order. When we say that men are taller than women, the words *on average* are implied. Pointing to the existence of your friend Rhonda, who really is quite tall, does not negate the statistical truth that, on average, men are still taller than women.

Some of the average differences between the sexes include that men have more "investigative" interests, while women have more "artistic" and "social" interests.[32] Men are also, on average, more interested in math, science, and engineering.[33] On tests, girls score higher in literacy, while boys score higher in math.[34] And although average intelligence is the same between boys and girls, the variability in intelligence is not: there are more boy geniuses, and more boy complete dullards, than there are girls in either category.[35]

One interesting piece of neuroscience reveals that, across several domains—including both emotional memory and spatial ability—women are better at details, men are better at "gist." This finding manifests, for instance, in the average man's superior ability to remember a route, and the average woman's superior ability to remember the location of the keys, the cup of coffee, the document in need of being signed.[36]

The differences between the sexes are found in babies, and across cultures, too—so this is not some weird WEIRD phenomenon. Given a choice, neonate girls spend more time looking at faces, while neonate boys spend more time looking at things.[37]

And across cultures, work is gendered early.[38] In an analysis of 185 cultures, in every culture studied, some tasks are always gendered in the same direction: iron smelting, the hunting of large marine mammals, metalworking—all of these are done by men only (in those cultures that do these things at all). More interesting is the tasks that are highly gendered across cultures, but for which some cultures curtail female involvement, while other cultures curtail male involvement. These include weaving, the preparation of skins, and the gathering of fuel, among others.[39] This suggests that there is a value in the division of labor, even when neither sex is inherently better at the task.

Consider, also, the Pueblo people, who have long been understood to be master ceramicists. It had been assumed, given contemporary patterns, that pottery making was exclusively the domain of women. In Chaco Canyon, however, in the Four Corners area of the American Southwest, a different story is emerging. When Chaco Canyon was a rapidly growing religious and political center one thousand years ago, the population was expanding, and with it, the demand for pottery. More and more vessels were needed to carry and store grain and water, so gender norms loosened, and men began to do this otherwise highly gendered work.[40]

What might we learn from these truths? We can learn that gender roles can be re-upped for modernity: some men will prefer hearth and home to a grueling career that is facilitated by having a spouse taking care of the domestic duties, and some women will prefer the latter. But many men and women, we argue, will prefer to be restricted to neither domain—without being slotted into preconceived roles, many people of both sexes will prefer a partner who is their equal, without being identical. We can learn from a more nuanced understanding of "gendered work" that traditionalist appeals to women not working outside of the home, or to men being dominant in economic and business matters, are regressive, without any nugget of necessity or truth. Historically, women and men have had division of labor, both in family units and in societies, but other than those tasks mandated by anatomy and physiology (gestation, lactation), there is little in the modern world that some women might not choose to do. Similarly, men are ever more welcome in traditionally female fields such as nursing and teaching, although we shouldn't expect parity there either. Different preferences lead to different choices. Pretending that we are identical, rather than ensuring that we are equal under the law, is a fool's game.

Sexual Strategies

It takes a lot to bring a baby into the world. While most readers of this book are likely living in a culture that assumes monogamy and biparental care, absent those constraints, men don't contribute much to the production of babies. It is also true that baby production does not end at birth, as once a baby has been successfully gestated for nine months, mother's milk may nourish that baby for six months, two years, or more, depending on cultural norms. Maternal investment in offspring is mandated, and high; paternal investment can be high as well, of course, but is negotiable. There are many humans alive now, and throughout history, who never met their fathers.

Across cultures, men and women report different preferences and priorities in what they are looking for in mates. In a now-classic cross-cultural study, mate preferences were investigated in thirty-seven cultures. In every culture studied, women were more interested in mates with high earning potential than were men. Furthermore, men were more interested in potential mates who are young, and physically attractive, than were women.[41]

Why would this be so?

Women who might get pregnant will have an easier time if that child has a father who contributes to the well-being of his child and mate. So women will be selected to prefer men with the capacity to earn. Because female fertility peaks early, and falls off far more steeply than male fertility does, men who might father children are more likely to be interested in youth and beauty in their mates—both of which can be understood to be proxies for fertility.

Furthermore, a woman who has given birth has certainty of maternity; she knows that she is the mother of that child. Certainty of paternity is far harder to come by, but important from a not-very-interesting-but-fundamental evolutionary perspective. Because fathers have never had certainty of paternity until recent technological advances have made it possible, the evolution of jealousy and mate guarding is far more prevalent in men than in women. Across cultures, men have tried to control the reproductive activities of women in such a way that they could increase their own certainty of paternity. Among the most patently divisive and destructive are menstrual huts that isolate women during menses (and so allow men to know when in their cycle women are) and female genital mutilation (which reduces or eradicates the possibility of sexual pleasure). Do not mistake our argument here: we are justifying no such control measures, but are merely investigating them with an evolutionary lens so that we can better understand them.

Another response to high uncertainty of paternity is this: there are a few cultures in which mothers' brothers act as the male role models for their sisters' children,[42] as de facto fathers under conditions in which they may not be sure of which children they have actually fathered. Among the Nayar of southwest India, for instance, wives and husbands do not live together, they share little beyond sexual activity, and women may have multiple husbands. With such uncertainty of paternity, fathers do no paternal care, but the mothers' brothers have both rights and responsibilities toward their nieces and nephews that are, to our WEIRD eyes, quite paternal. In general, though, a man who is tricked into raising someone else's children is ridiculed.

The last several paragraphs reflect established evolutionary theory.[43] Where it gets really interesting, in our view, is what it predicts about the reproductive—and therefore social—strategies that men and women deploy. We introduce these here, and explore their implications further in the next chapter.

Broadly speaking, there are three possible reproductive strategies:

1. Partner up and invest long term, reproductively, socially, and emotionally.
2. Force reproduction on an unwilling partner.
3. Force nobody, but also invest little beyond short-term sexual activity.

Women, being constrained by both gestation and lactation, and mate choice historically being in service of the production of children, have not had much flexibility in terms of strategy. Women were largely tied to strategy number one. Until very recently women preferred long-term partners to one-night stands, and they were far more likely than men to be sexually reticent ("coy").[44] Women have thus tended to engage in a long game, looking for a man with whom to partner, to co-parent, and to grow old together.

This is the best strategy for women if they want to leave their genetic mark in the next generation. Gestation and lactation are anatomically, physiologically mandated features of being a female mammal. Given that women are forced by nature to invest in their children, they have a better chance of successfully parenting if they have a partner who is invested as well.

This strategy—of looking for and staying with a long-term mate, with whom you raise children and make a life together—is also a possible reproductive strategy for men. All three of these strategies are open to men, but the first one is the male strategy that is best for society, best for children, best for women, and best for all but a handful of men—a position that we will defend in more detail in the next chapter. Strategy one is the long game, the emotionally invested strategy. This leaves two strategies that have historically been exclusively the domain of men.

Of these remaining possibilities that men have historically chosen, one is patently, morally reprehensible. Rape has allowed men to successfully reproduce, especially in times of war. Nobody will defend rape as honorable or desirable for individuals, or for society. Rape is reproductive strategy two.

The final alternative male strategy, however, is also not honorable or desirable for society, and yet has been encouraged for women by "sex-positive" activists as a sign of freedom and escape from puritanism. This third reproductive strategy is that of one-night stands. Of sex with strangers. Of sex without commitment or expectation. This strategy does not involve force—many women will often willingly sleep with a guy they just met. Such sexual liaisons do tend to come with some deceit, though, often by both parties. If women adopting some of the worst traits of men is our evidence of equality and freedom, we need to reinvestigate our values. This is reproductive strategy three: sex with neither force nor investment, the short game.

As women have increasingly engaged in reproductive strategy three, the short game, sex has become more mundane and easier to source. Contra the messaging of sex-positive feminists, engaging in this short game diminishes the sexual power of women. When people of both sexes are routinely seeking frivolous, no-emotional-connection sex, they are creating conditions in which everyone is behaving like men at their (second) worst. It's not as bad as rape, obviously. But it's not as good as strategy one, either. Society sliding toward this third reproductive strategy for both men and women is a variation on the Sucker's Folly—the tendency of concentrated short-term benefit (sexual pleasure) not only to obscure risk and long-term cost, but also to drive acceptance even when the net analysis is negative (in reducing the chances of finding love and all that flows from it).

Furthermore, women often don't intuit that there are two distinct male strategies (aside from rape), and so are often advertising toward men engaged in strategy three, the one that looks for one-night stands, while actually angling for men interested in strategy one. Hotness is a manifestation of sexual strategy three gaining primacy. Beauty, by comparison, is a manifestation of the first sexual strategy, the one that's in it for the long term. Hotness fades fast with reproductive potential. Beauty fades far more slowly.

We are due for a major renegotiation of the deal between the sexes. We cannot go back, and must not stay here.

Failures of Reductionism, Redux: Pornography

Finally, a cautionary note about pornography.

There is no such thing as "having sex." It's not like "watching Netflix," or "playing the guitar." Sex is interactional and emergent, such that "having sex" with person A is not the same as "having sex" with person B.

This is, once again, the mistake of reductionism, imagining that a proxy for a thing *is* the (whole) thing. Imagining that if we can count and record it, then we have counted and recorded the vital heart of the thing. Chemical imbalance *is* mental illness. Energy drinks *are* food. Porn *is* sex.

Wrong on all counts.

Of course people are fascinated by human sexuality. There is information, in watching others, about hazards and opportunities, both evolutionary and personal. Doing so triggers that third sexual strategy, though, the one that wants to get to the act of sex right away, the one in which it doesn't matter who it is that one is having sex with.

Just as jealousy varies by sex—men are more likely to be jealous of physical infidelity, women are more likely to be jealous of emotional infidelity[45]—so too do the target audiences of porn and erotica vary by sex. In general, women prefer erotica,[46] which brings with it a backstory, as it were. Porn, targeting that third sexual strategy, reduces human bodies to our constituent parts, and puts a premium on extreme sexual acts as a result of economic competition for attention. Among populations that have come of age on a steady diet of porn, women are far more likely to report being asked to engage in anal sex, strangling, and other violent "games" that are represented on-screen,[47] even though few real women want these things in life.

Porn, we posit, produces what we will call *sexual autism*.

Of course, we are using the word *autism* metaphorically. It goes without saying that we mean no offense to those who have been diagnosed with autism in a clinical sense, nor are we arguing that people with autism don't desire true connection, love, and relationships any less than all the rest of us. Here, we take the diagnostic criteria for autism and suggest that porn produces, in its adherents, something similar with regard to sexuality: incoming sensory data are of primary importance, and emotional and social communication is backgrounded, if considered at all.

Those who learn sexual behavior via porn tend to display repetitive behavior, and atypical sensitivity to sensory inputs. Sexual communication is difficult for them—probably because communication is a two-way street, and the other person cannot be fully predicted in advance, or controlled. People who have learned about sex through porn have difficulty developing, maintaining, and understanding sexual relationships; insist on inflexible adherence to routine; and show intense fixation on narrow interests. It is, in short, difficult for them to contend with novelty and surprise, with discovery ("I didn't know I could feel that way"), and with emergence ("I didn't know *we* could feel that way").

If the most complete human sexuality is, as we argue, an emergent property between whole individuals—bodies and brains, hearts and psyches— porn reduces sex to commodity, to acts, to mere bodies. Selecting from a narrow menu of options, sex learned from porn will be repetitive and inflexible, with a narrow focus on orgasm. Those who learn about sex from porn are likely to be insensitive to feedback from anything but their own body. Communication and feedback will not be priorities, nor perhaps understood as values at all. Relationships will be difficult to form, harder still to understand. Discovery and serendipity never happen when choosing from a menu. This is safer, in a way—while you risk not discovering the true highs of human relationship and connection, you are also protected from some of the true lows. Sex learned from porn can thus effectively flatten human sexuality. What of the world of emotional, deeply human discoveries that are possible with a richly connected sexuality? Without those, you might as well just be going for a walk.

The Corrective Lens

- **Avoid sex without commitment,** including sex for hire. Cheapening sex by seeking it everywhere and at all times renders it more difficult to form a stable bond with one individual, which is the best predictor of a relationship of equals, in which neither partner feels chronically subservient or undervalued. Instead, seek ecstasy and passion, which are more consistently discovered with someone whom you know well, and who knows you well.
- For straight women: **Do not succumb to social pressure to embrace easy sex.** If you aren't interested in sleeping with a guy within hours or days of meeting him, and he passes you by for someone who is—what have you lost? You have lost access to a guy dedicated to strategy three. You are better off looking for a good guy who is capable of, and interested in, refusing to act on his basest impulses.

- **Keep children far away from pornography.** Try to keep yourself away from it as well. The market should not be allowed to intrude on several things, including but not limited to love and sex, music, and humor.
- **Do not interfere with children's development by trying to block, pause, or radically alter their development.** Gender is the behavioral expression of sex, and so is both a product of evolution and more fluid than is sex. Childhood is a time of identity exploration and formation. Children's claims to be the sex that they are not should therefore not be indulged as anything beyond normal play and searching for boundaries. While actually intersex individuals are real and incredibly rare, and actually transgendered people are also real and very rare, much of modern "gender ideology" is dangerous and contagious,[48] and many of the interventions (hormonal, surgical) are not reversible.
- **Keep contaminants away from fetuses and children.** In several species of frogs, there is an established relationship between exposure to common environmental contaminants, like atrazine (an herbicide), and an increase in hermaphroditic individuals. While sex determination in frogs is different than in humans, we will not be surprised if it turns out that some of the modern confusion around sex and gender ends up attributable to widespread endocrine disrupters in our environment.[49]
- **Recognize that our differences contribute to our collective strength.** If we more highly valued work that women are more likely to be drawn to (e.g., teaching, social work, nursing), perhaps we could stop demanding equal representation of men and women in fields that women are simply not as likely to be interested in. Recognizing that we are, on average, different is the critical first step to building a society in which all opportunities are truly open to everyone. Equal opportunity is an honorable goal in step with reality, whereas aiming for equal outcome—in which every occupation, from day-care workers to garbage collectors, has equal representation between the sexes—will disappoint everyone involved.

Chapter 8

Parenthood and Relationship

THE INDIVIDUAL—THE SELF WITH A BODY AND BRAIN, WITH LEGS AND blood and thoughts and emotions—is a complicated phenomenon, one that we have been focused on so far. When you bring individuals together, in relationship, the complexity grows exponentially. In many animals, the interplay of individuals and all their complexities has resulted in a force of nearly transcendent power: love. In humans in particular, love is profound.

All love has a common origin story, although its forms can feel so different—the love for a child, for a spouse, for a cause. All are beautiful, and can be disruptive of normal life. Our persistence as a species has been possible, in part, because of love. This raises the question: What is love?

Love is a state of the emotional mind that causes one to prioritize someone or something external as an extension of self. That's it. Love, the genuine article, is a matter of intimate inclusion. When it is real, there are few forces more powerful.

Love evolved first between mother and child, but then spread its wings, expanded its scope. Soon enough, adults reliably experienced love between partners, and then other forms of love began to blossom—between fathers and children, grandparents and grandchildren, among siblings. Love then found a place between friends and between soldiers, between those who shared intense experiences, good or bad. Much human mythology is centered on inducing people to extend their concept of self, and to shape the in-group to which the concept applies—the Good Samaritan story reveals the capacity for love even between those who are supposed to be enemies.

Eventually, love evolves to include abstractions—love of country and love of God, love of honor and service, truth and justice.

Love as we experience it first evolved nearly two hundred million years ago, when mammals diverged from reptiles. As with the evolution of sex, the egg is foundational to our understanding of the evolution of love. The most recent common ancestor of mammals and reptiles laid eggs. In egg-laying species, the egg must contain sufficient nutrition to feed the embryo through hatching. And in species in which both parents walk away and never meet or care for their young, the hatchling must also be capable of immediately feeding itself. A mother can stack the deck in her brood's favor: a butterfly can lay her eggs on a plant her caterpillars are equipped to eat; a wasp can lay eggs inside the paralyzed body of a spider, which her young eat their way out of; an octopus can die in contact with her hatching eggs, thereby handing over the nutrition of her own body to her hungry offspring. Without parental care, though, the hatchling is on its own.

The first mammals were egg layers, and eggs do not require love, although in many species they benefit from parental vigilance. But the five extant species of egg-laying mammals—four species of echidnas, and the duck-billed platypus—are substantively different from all other egg-laying species. Mammals, even the ones that lay eggs, make milk. Early on, it was a crude operation: modified sweat glands secreted nutritious fluid that was lapped up off the surface of the mother's skin. Later, a more elegant solution for delivery evolved: the nipple. In all mammals, nippled or not, milk solves a problem.

A mammal mother can leave her babies in a safe place while she forages, freeing her from having to provide all their food in advance or having to shuttle morsels back to her burrow. Milk also allows the baby's food to be chemically and nutritionally adjusted in various ways that facilitate development. That's it, at first. Mother's milk is just one of many evolutionary answers to questions of nutrition and immunity. And it is the gateway to much more.

Once milk glands become integral to offspring maturation, babies are guaranteed to meet and spend time with their mothers. Until recently in human history, direct mother-baby contact was the only way that transfer could take place. Love isn't required for this. Mothers and babies could just be hardwired to do their part with no emotional involvement. Like complex sociality and extended childhoods, however, emotional involvement is adaptive. Add to this the fact that pretty much any predator big enough to consume the babies will see them as a delicacy—defenseless, tender, and probably free of pathogens. They are practically the perfect food. That means that a mammal mother will often be confronted with a question: How much risk should she take to protect her babies when they are threatened?

Each mother and each situation are different, and the calculation requires that the mother has information on a number of things. How much of her reproductive life remains ahead, and how much is behind her? How dangerous is the predator she is facing, and how well equipped is she for battle? If she dies saving one offspring, will she doom others to starvation? When all considerations are taken into account, there is, in effect, an ultimate calculation: Is her fitness more likely to be enhanced by a given confrontation, or diminished? All else being equal, lineages in which mothers calculate more accurately will outcompete more crudely informed rivals, and calculations will be improved and adjusted by this process over time.

Of course, animals don't do calculations in any explicit sense, nor do they have access to data on reproductive life span, hazards, or opportunities. What they do have is an internal architecture that has been tuned by selection to intuit these things and adjust behavior accordingly. The language in which these intuitive calculations manifest—the way they motivate behaviors—is through emotions. Love is a powerful amalgam of those emotions.

In this chapter, we will explore the ways that love has evolved to motivate our family dynamics, affected how and with whom we mate, how we age, why we grieve, and more.

Parental Care: Moms, Dads, and Others

All mammals are cared for by their mothers, and mother's love, we argue, is the most ancient and fundamental form of love. All true love is an elaboration on this concept. But mammals are not the only creatures in which love has evolved. The other place where this pattern flourishes—a whole separate evolution of it—is in birds.

There are many bird species in which parents and offspring never meet. Bush turkeys lay eggs in a mound from which chicks hatch and disperse, already self-sufficient. Cuckoos, cowbirds, and other "brood parasites" lay eggs in nests being tended by birds of other species, and in each case the hatchlings must be preprogrammed, as their unwitting stepparents are in no better position to convey good lessons to baby cuckoos than you, dear reader, would be to teach a marmoset how to navigate a life in the trees. These prewired birds are exceptions, however. In the vast majority of bird species, parents actively tend their young, and these nurturing parents face exactly the same considerations about fitness and risk as mammal mothers do. Likely you have seen smaller birds mobbing larger, predatory birds to drive them away from their nest. That's love for you.

In all mammals, and in that majority of birds in which parental care is the rule, offspring are fed and protected by their parents. This leaves the young developmentally free to evolve toward helplessness—they don't have to defend or feed themselves if a parent is there to do that job for them.

Helplessness of hatchlings and newborns—*altriciality*—is not itself an asset, but this helplessness opens the door to extraordinary things. Major programming of the brain can take place through cultural transmission when offspring are in close contact with their parents. This is far faster than genetic change, and allows not only for rapid behavioral evolution, but also for tailoring of behavioral patterns to the local environment—physical, chemical, biological, and social.

In birds and mammals, altriciality[1] is the downside of behavioral flexibility, which *is* an asset. Behavioral flexibility—*plasticity*, which we will return to in the next chapter—emerges in organisms that are not fully programmed by the genome. Painting with a broad brush, plasticity increases both in species that have interaction between the generations and as helplessness in hatchlings and newborns increases.

Among animal species with parental care, mom is the most common primary caregiver, although exceptions do exist: jacanas and seahorses reveal that sex-role reversal in caregiving under some ecological conditions can and does evolve. Paternal care absent maternal care is thus rare, but not unheard of. Biparental care, in which both parents actually help raise the offspring, is more common. Biparental care is typical when monogamy is the norm, in organisms as varied as swans and arctic terns, titi monkeys and gibbons. Finally, in many species, from superb fairywrens to meerkats, siblings and even unrelated acquaintances help raise babies, a system known as cooperative breeding.

Callitrichids are a clade of New World Monkeys that include marmosets and tamarins, in which cooperative breeding is common. Mothers tend to have twins, and nursing and foraging take up the entirety of the mother's time and energy budget. But babies require constant carrying, and young juveniles require constant vigilance, such that they don't fall out of the jungle trees in which they live. Who will do these things—carry the babies, watch the juveniles—if not the mother? While only mothers can nurse, in many species of callitrichids, all other care of the young is done by members of the troop other than the mother—the father, sometimes his brother, older siblings, and temporarily nonreproductive females who join the troop on the chance that they may one day inherit it. Similarly, cooperative breeding is on full display in naked mole rat pups, which, after being weaned, are taken care of by workers rather than parents.

What causes the transition from independent breeding, in which individuals are simply pursuing selfish interests, to cooperative breeding, a system of greater complexity and collaboration? In part, cooperative breeding, which is on full display in many human societies, is most likely to evolve when rates of promiscuity are low[2] and resources are distributed across the landscape such that any given individual cannot monopolize them. The monopolization of resources opens the door to the monopolization of mates—indeed, the distribution of resources in space and time has far-reaching effects on mating systems.[3]

Mating Systems

Imagine a mated pair of swans, swimming together. The male may be slightly larger, but they are so similar as to be difficult to tell apart.[4] Across monogamous species, males and females closely resemble each other in color, size, and form. Compare this to elephant seals, which are decidedly polygynous—a single male can monopolize the reproductive activity of dozens of females. Males have outsize noses and are more than three times the size of females. Sexual size dimorphism is a strong predictor of polygyny across vertebrate species.

Humans are far closer to swans than to elephant seals when it comes to sexual size dimorphism, but we are no swans. Men are, on average, about 15 percent larger than females,[5] and significantly stronger. This tells us that our ancestors were at least somewhat polygynous or promiscuous.

The polygyny in our evolutionary past shouldn't surprise us—none of the other extant great ape species are monogamous. But *Homo sapiens* have apparently been evolving in the direction of monogamy since our divergence from chimps and bonobos—we are less sexually dimorphic than either of those species. And while the majority of human cultures have been polygynous at some point in time, the majority of people alive today belong to cultures in which monogamy is the norm.

Monogamy is fragile, and easily and often breaks down into polygyny in mammals. Despite this, monogamy is the superior system.

Types of Mating Systems

Mating system refers to the number of mates members of each sex typically have. Broadly, the types are:

- **Monogamy:** individuals of both sexes have just one partner at a time
- **Polygamy:** individuals of one sex have just one reproductive partner, but individuals of the other sex have multiple partners. Subtypes include:
  - **Polygyny:** (poly—many, gyn—female): One male and multiple females
  - **Polyandry:** (poly—many, andr—male): One female and multiple males.
- **Promiscuity:** members of both sexes have multiple partners (in humans this is sometimes called polyamory)

To defend the bold claim that monogamy is the superior system, let us start here: Monogamy is the mating system with the greatest potential for cooperation and fairness, beginning with child-rearing. In primates, monogamy is also correlated with the largest relative brain size.[6] Across the biota, females are the limiting sex, which allows them to be choosy about their partners. In a polygynous system, sexual partners abound for females, but they are in short supply for males and—absent an intent to invest beyond the act of sex—males tend to have incredibly low standards for sex partners. If a female does not show obvious signs of a communicable disease, a male is likely to accept just about any willing female, even if she is not of precisely the right species. A tiny chance of producing offspring, even hybrid offspring, is evolutionarily better than no chance at all.

Absent monogamy, this is what sexuality reduces to: females burdened with the entire chore of reproduction, and undiscerning males always on the make.

When monogamy does arise—when both females and males are pursuing what we called strategy one in the previous chapter—males become more like females both in their perspective on sex and in their morphology. Because monogamous males select a female and forgo sexual opportunities with others, they have as much reason as females to be choosy about sex partners. Males being choosy in this way reduces their tendency toward violence. They may yet fight over access to the best females, but no longer need aspire to the acquisition and defense of "harems," which are closely associated with aggression and physical weaponry like antlers and piercing teeth. If we compare gibbons, which are monogamous, with baboons, which are not, we see that baboons have marked sexual size dimorphism and enlarged canines. Polygyny—which is associated with strategies two and three from the previous chapter—leads inexorably both to male-male violence and to the morphology that enables that violence.

Monogamy also creates a system in which nearly everyone has a mate, as sex ratios tend to be one-to-one within populations, regardless of mating system. This prevents the accumulation of sexually frustrated males for whom violence may be the only path to reproduction, either through the overthrow of harem owners—as in lions and elephant seals—or through rape—as in ducks and dolphins. We will return shortly to some of the profound implications of monogamy for human societies.

Birds and mammals, for all of our separately evolved similarities, differ markedly with regard to mating system. Few mammal species are monogamous, but most bird species are at least somewhat so. That is to say, most birds have extended periods of sexual exclusivity—one male, one female. Some pairings last for a breeding season, some couples mate for life. Why the difference?

All bird species lay eggs. Bird eggs, strange as this will sound, are a powerful antidote to male sexual jealousy. That's because avian eggs are fertilized right before the shell goes on, and shortly before they are laid. As such, a male bird need only guard a female against his rivals for a brief period of time before and after mating in order to be certain that he is the genetic father of her brood.

Live-bearing mammals, by contrast, have a long period between fertilization and birth, and most males cannot therefore be sure that a female with whom they mated did not also mate with another male around the time of conception. Males without "certainty of paternity" are unlikely to stick around to pair-bond with a female and help raise the kids. Male birds tend to have high certainty of paternity, but male mammals are rarely certain at all. As a result, male mammals tend to abandon mates and offspring when—if they could just be confident of their paternity—selection would clearly favor their sticking around to help. Mammals have a harder time evolving stable monogamy even though it is, in most regards, the superior mating system.

Once pair-bonded, a male is faced with a choice. He can merely guard his chosen female against rivals, or he can contribute in some way to the provisioning and care of their offspring. Paternal care is not universal in monogamous species, but it is common. Paternal care increases the chances that offspring will live to be reproductively viable, and also increases the number of viable offspring that can be produced—both of which contribute to the fitness of male and female alike.

Monogamy thus expands the reach of love, beyond that between mother and child to that between pair-bonded mates, and often to between father and child as well. Friendship can also be facilitated by monogamy. Jackdaws, which are relatives of crows, form lifelong pair-bonds, and as they fledge, they form friendships with jackdaws of similar age, gifting one another with food, which helps them form strong affiliative bonds.[7]

Pair-bonding also creates an opportunity for a useful division of labor. In a uniparental system, the single parent—usually the mother—simply has to do it all; pair-bonding stands to cut her work in half. Among western sandpipers, mothers incubate eggs in their arctic breeding grounds at night, and fathers take over during the day.[8] Monogamous freshwater fish known as Midas cichlids have fathers that focus on territorial protection and mothers that focus on nurturing their young.[9] Pygmy marmosets, whose mothers spend all their time procuring food sufficient to meet their own and their babies' needs, have fathers that do all the remaining infant care.[10]

In humans, there seems to have been a positive feedback loop: as babies got more and more helpless, with longer and longer childhoods, the bond between co-parents got tighter and tighter. Love is a manifestation of the tightness of that bond.

As families have evolved, so too has love between siblings. The flip side of such love is sibling rivalry, which is a strong force across all of those species in which siblings meet one another. When parents are pair-bonded, their offspring will be genetic full siblings. By contrast, in species in which mothers do all the care—and males engage wholly in sexual strategy two or three—offspring may be half sibs, so only half as related to one another. From a purely genetic standpoint, full sibs thus have twice the genetic basis for cooperation as do half sibs. Given that monogamy is the route to having full sibs, monogamy thus enhances cooperation between offspring and reduces the tendency of intra-brood conflict. An extreme case of such cooperation is known in naked mole rats, which have evolved eusociality reminiscent of that seen in ants, bees, wasps, and termites.

In mammals, the relatedness of siblings has another rather bizarre implication. Illnesses faced by mothers during pregnancy are often the result of a conflict of interest between mother and fetus—an often gentle but real tug-of-war for resources.[11] A mother has a strong interest in dividing resources between her various offspring over her entire reproductive life span, whereas a developing fetus—which has hormonal access to its mother's bloodstream and is only 50 percent related to its mother—has an interest in taking more than its share, so long as its take doesn't put its mother's life at risk. That conflict of interest is mitigated in monogamous populations, in which the fetus has double the interest in the survival of future full siblings compared to populations in which future siblings will be sired by different fathers. From the perspective of a father in a nonmonogamous species, another way to put this is that males take advantage of the maternal behavior of females, and their genes continue this parasitism through pregnancy.

Implications of Monogamy in Humans

Sex, gender, relationship, and mating systems: four topics intimately intertwined and shot through with complexity and significance. There is little more central to the human experience. We discussed these themes in the previous chapter, but here we return to them with a bit more context and nuance.

Mating systems shift as ecological conditions change.[12] When resources are in surplus, monogamy is likely favored. It provides a clear advantage at the lineage/population level by bringing all able adults into child-rearing, allowing the population to capture resources at the fastest rate. When a population reaches carrying capacity, though, and zero-sum dynamics are once again in play, the incentives of high-ranking men tend to shift in the direction of polygyny. Male-male competition becomes a driving force, as men with wealth and power seek to dominate the reproductive output of multiple women. In WEIRD countries, this pattern can be obscured, because when a man leaves his family to have a new one with a second (usually younger) woman, we call him "serially monogamous," but in reality, this is clearly a form of polygyny.

When polygyny is on the rise in a society, we see an increase in sexually frustrated young men who are willing to take big risks for the possibility of gaining a mate. The lineages of the relatively few powerful males who benefit from polygyny also benefit by arming frustrated young men and sending them abroad with dreams of coming back with a bride or as a marriageable war hero. The possibility of gaining territory and treasure through military adventurism has clear evolutionary implications—it expands the resources available to the conquering population and thereby increases its size. Such military adventurism is also a "transfer of resource" frontier, which, as we discuss in the final chapter, is a form of theft.

The costs of polygyny can still be obscure, though, so here are some more, from research that models the effects of mating system (monogamy versus polygyny) on fertility and economic status, and compares that model with existing empirical data. All else being equal, people in monogamous cultures experience lower birth rates, and higher socioeconomic status, than those in polygynous ones. Furthermore, the age difference between spouses is smaller.[13] In part, this likely reflects a move away from viewing women and girls as commodities, which strongly polygynous cultures often do.

Polygyny is often conflated with a fantasy of promiscuity in which there is an increase in sexual and reproductive freedom without negative consequence. The sexual revolution would seem to bear this out, but it is an illusion for two reasons. First, absent birth control, promiscuity is good for men, who procreate almost without cost, and dangerous to women, to whom the entire burden of child-rearing gets shifted. Second, promiscuity tends to break down into polygyny as powerful men discover that they are in a position to demand exclusivity from multiple reproductive partners, either in sequence or in tandem.

Birth control changes the entire reproductive equation for both better and worse, giving women the ability to avoid that shift, but also spectacularly reducing the apparent value of women to men, and making men more reluctant to commit.

Prior to birth control, women (and their close kin) guarded their reproductive capacity with tremendous diligence. Because human babies are so hard to raise, a woman who could insist on help from a man would be foolish to forgo it. In such a world, men had great incentive to impress women to whom they would commit. Our modern arrangement gives women a lot more freedom to enjoy sex without the risk of becoming single parents, but it also radically undercuts their bargaining position with respect to long-term commitment—especially if they engage in reproductive strategy three, the short game.

Men might seem to be getting the better part of this deal—and indeed with respect to physical pleasure, easy access to sex is a prize from which most men cannot look away. But as we discussed in the previous chapter, the sex in question is low stakes, rewarding in the moment but meaningless in the long term. Men may physically feel as though they have been judged sexually worthy by large numbers of women, but subconsciously they know that the bar for such acceptance has been lowered so far as to be meaningless. Yes, it is sex, but it is junk sex, formulaic and without depth.

Women may consciously want to enjoy sex without commitment, but they are wired to fall in love with the men they bed because sex, babies, and commitment are evolutionarily inextricably linked for women. Sex and orgasm trigger the release of oxytocin in females, which promotes bonding. The situation is similar for males, except that it is not oxytocin but vasopressin that does the job (the role of these hormones in human sexual and social behavior has been well studied, but their relationship to pair-bonding is even more extensively documented in monogamous prairie voles, whence much of what we know about these hormones' role in bonding comes).[14]

Given the inherent imbalance of reproductive investment between men and women, though, we predict that the system is even more complex than this, at least where men are concerned. Consider that if you are a man engaging in reproductive strategy two or three, it does not benefit you to fall in love. While reprehensible, these strategies have been effective for men throughout history, so selection has surely facilitated them under some circumstances. Such strategies tend to involve sex quickly upon meeting a sex partner, so we predict that vasopressin will not be released, or will be released in far lower amounts, in men who are having sex with women they just met. The pair-bond-facilitating vasopressin, we argue, will be most present in men when they have sex with women they have known for a longer time. If true, what this would mean for women is that sleeping with a guy you're really into on the first or second date will actu-ally *decrease* the likelihood that he will fall in love with you.

In this world of low-stakes sex, many men have lost one primary in-centive to achieve in myriad realms, and have become mercurial about the very commitments that many view as life's greatest source of meaning. Women have been freed from some constraints, to be sure, but released into a world of adolescent sexuality, an endless game of shallow partner-ships without purpose.

Who wins from this arrangement? Two categories of men benefit: the wealthy and powerful who are in a position to have multiple partners, and those who delight in pretending to be interested in commitment in order to bed women in whom they then invest nothing. Some men in both of those categories also propose sexual quid pro quo with women trying to advance their careers, a position that women can find impossible to re-cover from without damage. Somehow we have replaced a deeply flawed system of mating and dating with one perfectly positioned to transfer all the spoils to kings and cads.

Things are even worse where sex ratios have been distorted from their usual balance by war or other forces. When eligible men are scarce, they are in high sexual demand. This puts women in a quandary. How does one get a man's attention when everyone else is attempting the same? Sex sells, and scarce men find many sexual opportunities and little reason to commit. The result is that women who want families in such environments often find themselves having to accept the cost of single parenthood.

This is exactly what has happened among men for whom a lack of economic opportunity and underfunding of schools have led to elevated mortality, crime, and imprisonment. In the United States, a large fraction of black men have been forced into such a situation. Men who avoid these bad fates find themselves in high sexual demand and tend to play the field, leaving many black women to raise families without a committed partner. Many in the ruling class have long pretended that this pattern derives from some imagined moral failing among black people, when instead it plainly emerges from demographics and game theory that would produce the same pattern in any population faced with similar conditions.

So we are all caught in a bind. Committed relationships are a good thing, so valuable to the rearing of healthy children. Yet if women in the modern mating and dating scene don't accept casual sex as normal, they're often ignored. If they do embrace casual sex, they often unwittingly trigger fear of commitment. Men could be seen as profiting at the expense of women in this situation, which is partly true, but their windfall is mostly an illusion. Yes, men are built to find commitment-less sex rewarding, but they are also built to value loving partnerships. Casual sex is disrupting that.

Male and female are complementary states, and there is a healthy natural tension between them. There is, of course, plenty to say about the implications and evolution of homosexuality, in both humans and other species. While we have no room in this book for such an analysis, we will offer the short tease here that while lesbians and gay men are both homosexual in that they are attracted to individuals of the same sex, the differences between the two, in terms of both evolutionary origins and how the relationships tend to play out for those in them, are large, and consistent with the differences between women and men that we have laid out in the previous two chapters. Furthermore, both forms of homosexuality—female same-sex attracted and male same-sex attracted—are adaptations.

That said, heterosexuality is the norm, and not for socially constructed reasons. Among straight men and women: if women conclude that in order to be equal they must behave like men when it comes to sex, then the system breaks down into one in which everyone behaves like men at their most adolescent. In spite of its stodgy reputation, monogamy is the best mating system. It creates more competent adults, reduces the tendency to engage in violence and warfare, and fosters cooperative impulses.

Elders and Senescence

Of all the systems essential to the functioning of humanity, parenting may well be the most compromised by hyper-novelty of the 21st century. To see why, permit us to go down a path that is related, but that may not seem so at first. Consider the much-ballyhooed quest to stop human aging. From fantastical historical accounts of a Fountain of Youth, reported from Herodotus to Ponce de León, to rich moderns who freeze their brains in the hopes of being resurrected once we have "cured" senescence, living forever has long been a desire of some mortals.

All else being equal, extrapolating from rates of mortality at reproductive maturity, if the human body could be made to simply maintain and repair itself fully, half of all people would live to a staggering twelve hundred years of age.[15] You might wonder, then: How hard a problem could it possibly be? The answer is, much harder than it seems.

Imagine, though, for the purposes of this chapter, that the senescence of the body turned out to be an easier problem to solve than curing cancer or the common cold. Imagine, further, that despite major differences between brain tissue and the rest of the soma, senescence of the brain could somehow be cured, too. What about the mind?

That may seem like a distinction without a difference, but brain and mind are not synonyms—the mind is a product of the brain. Brain is hardware; mind is software. A perfectly functional piece of hardware is of little value if the software and the files are seriously corrupted. If we fixed the brain without somehow retooling the mind—even if every physical pathology of the brain could be banished—most of the centuries of extra life would be wasted in a nightmare scenario of senile minds hobbling youthful bodies.

Human minds have a capacity. In order not to fill up with ephemera, we must forget almost everything, a process that leaves us unreliable witnesses to the events of our own lives. At very advanced ages even the most cogent among us have decidedly fragmented cognition. How much more destructive would the process be if it had to budget for centuries?

Humans are the longest-lived terrestrial mammals. Many of us have more than eight decades of useful life on this planet, but that life span pales in comparison to the deep evolutionary solution for the problem of aging, by which we *do* approach immortality: our descendants.

We spend decades acquiring skills and knowledge about how to function effectively in the world, but this hard-won programming is trapped within bodies that soon succumb to the hostile forces of nature. Were we neurologically hardwired by our genomes, we would be born knowing how to be adults, and we could get there directly. Even among organisms that do have parental care, there is often considerable preprogramming. A horse, for instance, is on its feet within minutes of its birth, sensing the world and moving about in nearly adult fashion, and we could do the same if our lives were as straightforward as those of equids. That is not to say that horses have nothing to learn: they must learn both their social roles in their herd and a map of the hazards and opportunities in their environment. But the basic parameters of being a wild horse are similar, no matter the environment. This allows horses to be precocial—highly capable from birth.

For humans it is the opposite. The human niche is niche switching. Our niches transform radically, sometimes over remarkably short distances. Consider again the difference between populations of arctic hunters who hunt big game a few kilometers inland and those who specialize in aquatic mammals at the coast. Those specialties require radically different skill sets, and would not be possible if each individual had to discover the secrets to hunting effectively on their own. The solution to this conundrum is so familiar to us that we rarely think to marvel at it. Recall the Omega principle, which points out that any expensive and long-lasting cultural traits should be presumed to be adaptive, and that adaptive elements of culture are not independent of genes.

Elders pass on the durable knowledge and wisdom by this second mode of inheritance—culture. Because this second mode is cognitive rather than genetic, and culture changes faster than genes, the niches we exploit can change at a staggering rate. This plasticity allows bands of humans to function as coherent bodies, with tasks divided as if into separate organs. These bodies in turn can fan out across physical landscapes, tailoring behavior slightly to suit a particular hilltop, or shifting radically to accommodate a brand-new food source.

In the shale hills of central Portugal, where farming is physically difficult, parents do the backbreaking work of growing and harvesting food, and grandparents raise the children.[16] In Portugal, then, menopause may mark the beginning of active child-rearing. Menopause, which is nearly unique to humans, is not the end of vitality for women. It is the end of direct reproduction, but the ability to provide wisdom and care to the young—to children and to grandchildren—after the risk of continuing to produce new babies is past, can be considered a great gift.

In the islands of the Andaman Sea, the Moken people have lives that are deeply connected to the ocean, and their stories of tsunamis being evidence of angry gods saved whole villages during the 2004 Boxing Day tsunami. In the aftermath of that massive tragedy, one old man reported not on the gods, but on his own direct experience, having lived through something comparable in Myanmar when he was a child.[17] Such wisdom does not become dated, and the older our elders, the more likely they have direct experience of other world-changing events, and know what to do in their face.

The wisdom of elders is ancient and necessary in human history, *and* there is deep value in being skeptical of the wisdom of elders, when that wisdom is out of place, or of the wrong time. Parents have an overriding interest in their offspring being highly effective in whatever environment they will inhabit. If the mind's software needs updating and the young are in a position to accomplish it, it is in everybody's interest that the antiquated paradigm be replaced. This explains why, for healthy parents, seeing oneself in one's children is some mixture of rewarding and jarring, but seeing them thrive is a thrill and a relief.

And this brings us back to children as the solution to the problem of human aging: Upload the useful subset of skills, memories, and wisdom into young bodies that have been wired to facilitate, augment, and amend this cognitive package as needed. It is entirely natural to want to live a long life, and to see one's descendants well positioned for the future. Many believe that as individuals we are entitled to more, that we personally must be preserved, but this desire is in error. Such preservation would interrupt the primary mechanism by which humans innovate and keep pace with change. We reject this ancient mechanism at our peril.

Love across Species

In pursuing lighter fare than considering that children, not immortality, are the antidote to aging, let us ask: We love our pets, but do they love us?

Humans have domesticated dozens of species of animals across the globe, mostly to provide food or do labor for us. Some of those relationships that began as purely functional—cats as mousers, dogs as protection—have since become cross-species friendships. Cats have befriended us for far less time, and remain more wild than dogs, more of their original selves, although they bond tightly to humans under the right circumstances. Even before we began to farm, though, dogs were by many of our sides, becoming domesticated. As hunter-gatherers, some of us already had dog friends.[18]

Dogs are in many ways a human construct. We have co-evolved with them for so long that they are now attuned to human behavior, language, and emotion. Perhaps you could argue, then, that humans are also partially a canine construct.

Does your pet love you? Of course your pet loves you. (Qualifier: your pet can love you if it's a mammal or one of a few clades of birds, like a parrot. If your pet is a gecko or a python or a goldfish, your pet is probably incapable of love.) Love develops for every evolutionary pairing that requires devotion. We love our pets, and our pets love us. Dogs, in particular, are love generators who hang out with you and help you know that you're not alone. Dog is love, unmoored.

Watch how cats and dogs engage with each other, and with us. They don't use language to convey meaning and emotion, but convey it they do. You have no reason to doubt that your dog is disappointed when you stop throwing the ball for him, or that your cat would prefer that you stay seated with her in your lap. We name our emotions—love, fear, grief—and when we attribute those words to animals, we may be accused of anthropomorphizing. As Frans de Waal, who has spent a lifetime studying emotion in animals, points out, this argument is rooted in assumptions of humans being not just exceptional, but wholly different from the other animals with which we share ancestors.[19] We need to be careful in how we attribute emotion and intention to other animals—as we should within our own species as well—but there can be no doubt that many other species plan and grieve, love and reflect.

In our interactions with our pets, we read their cues without language. It is helpful, too, in your interactions with people, to turn the sound down sometimes. Be an animal behaviorist—or just act, sometimes, like a human before language evolved. We often use language to cover how we actually feel, to deceive, to throw off the scent of what is actually going on. When you watch people, especially strangers from a distance, it's relatively easy to read the emotion of the situation. Pay attention to people's behavior, not the stories that people tell about their behavior. That's what your dog is doing. Your dog doesn't buy your cover story—although he is likely to forgive you for your foibles.

Grief

In his *Metamorphoses*, Ovid writes of the old married couple Baucis and Philemon. They have been poor all of their lives, but generous with what little they have. The gods recognize them for their righteousness and ask what they would like most in the world. Baucis and Philemon ask that when death draws near, they might die together, so that neither shall see the other die, nor be left behind. The gods make it so. The old lovers become trees—an oak and a linden—which intertwine their branches with one another as they grow.

Absent interference from Zeus, the only way to avoid grief is to live a life without love. Grief has evolved multiple times across species, always in highly social organisms with parental care. The grief of chimps is presumably the same, at its root, as human grief. The grief of dogs has distinct origins, though, as the most recent common ancestor that we share with dogs was a nondescript little mammal with little to no social structure. Perhaps the most famous story of how grief manifests in dogs is that of Hachikō, a handsome Akita who was born in Japan in 1923. Hachikō was taken in by Hidesaburō Ueno, a professor of agriculture who commuted to work by train. Every day, when Ueno was due back, Hachikō met him at the train station, and they walked home together. One day, Ueno did not return, having died of a cerebral hemorrhage while lecturing. Hachikō, however, continued to go to the train station every day and wait for his master, for nearly ten years, until he himself died.

The grief of canids—wolves and domestic dogs—evolved separately from our own. Elephants also grieve, as do killer whales. In all of these cases, grief is independently evolved, yet deeply similar: an extreme emotional response to the death of an intimate, unpredictable in its length and manifestation.

Modern approaches to loss and to grief tend to overemphasize metrics and logistics (How many years had he been sick? How do I obtain a death certificate, close bank accounts, cancel appointments?) and spend too little on meaning and narrative (What did he bring to us? How are we better for him having lived?). We often don't want to see the body, or sit with it at all. Death is at a remove, and this particular hyper-novel situation, in which it is our choice to not confront the corpse of a loved one, can render us more confused in the aftermath of death.

Grief is us recalibrating our brains for a world without one of its central pieces. We must reformulate our understanding, as we are no longer able to go to that person (or animal) for words of wisdom or comfort, but we are still able to think back, to learn from, and to take comfort in the relationship that can no longer grow, but can still be remembered. We don't want to believe in their permanent absence, and so our brains create fictions, ghosts: Was that him turning the corner at the café we used to frequent? Surely that was her—I know her hair, her jacket—getting on the train.

Grief is the downside of high-bandwidth interdependence. Grief is the downside of love.

All too often, moderns try to protect children from grief. For instance, we have known parents who would not allow their children to attend their grandparents' funerals, for fear that it would scare them, or harm them. This fear and anxiousness in raising children, in turn, creates fearful and anxious children. In the next chapter, we explore childhood, and how to raise children who are independent, exploratory, and full of love.

The Corrective Lens

- **Take time to grieve in a way that feels right to you.** In the middle of deepest, earliest grief, sometimes you will be joyous, and sometimes you will not be thinking about the one whom you have lost. It ebbs and flows, loses some of its power over time, but never disappears entirely. No matter what, honor your memories and your inclinations.

- **Spend time with the body of your loved one after they die.** Those who have lost loved ones to situations from which their bodies could not be recovered often suffer from prolonged periods of grief. When we view our dead, sit with them, and talk with them, we set a foundation upon which our grief, our neural recalibration, can be moored.

- **Turn the sound down on your conversations, and watch the actions.** Act like an animal behaviorist, especially when interpreting interactions between you and your romantic partner. Much can be learned about the actual emotions in play when you stop listening to words and start watching behavior.

- **Be an animal behaviorist on your own emotional state** as well. And recognize that if you have contempt or disgust or persistent anger for someone with whom you are in a relationship, these feelings are incompatible with love.

- **Avoid dating apps,** if you can. In a world of billions, in which city dwellers interact anonymously with people every day, dating apps may be a good way to sort through a nearly infinite number of choices. The risks are many, though, including that seeing so many possible partners may make you less interested in going deep with any individual. In such a sea of opportunity, it is also possible to become a perfectionist—to get the sense that the "perfect one" must be out there, if you just keep swiping long enough. Best to have real-life interactions early and often in any relationship that you think might be worth developing.

- **Encourage alloparenting of your children**—by grandparents, older siblings, friends. If you have only adult women, or only adult men, in your household, an alloparent of the opposite sex stands to be particularly beneficial for your children.

- **Breastfeed your infants**, if you can. Adults who were breastfed have better-formed palates and better-aligned teeth compared to those who were bottle-fed;[20] and breast milk has in it all manner of nutrients and information that we do not understand. It may, for instance, contain cues with which the infant entrains his sleep-wake cycle. Thus, if you do breastfeed, and also pump milk to feed the baby at other times, feeding your baby milk that was pumped at the same time of day as it currently is could be helpful in getting your baby to sleep when you want him to. Put another way: beware Chesterton's breast milk.[21]

Chapter 9

Childhood

CHILDHOOD IS A TIME OF EXPLORATION. IT IS A TIME TO LEARN RULES, to break rules, and to make new rules.

When our older son, Zack, was five years old, he innovated a new mode of locomotion down stairs, which involved a large rubber ball and a mattress. It worked well, until it didn't. His arm break required the surgical insertion of metal pins to stabilize the growth plate of his humerus, and another surgery six weeks later to remove them, but he healed beautifully, and innovated with more wisdom going forward.

A young orangutan following his mother through the trees may whimper and call for her upon reaching a gap that is too large for him to cross. She will return and bridge the gap, allowing him both to cross and to see how it is done.[1]

Young ravens spend years in large social groups after they become independent from their parents, and before they form pair-bonds with long-term partners. Alliances form during this time, but conflicts also arise, and those ravens that figure out how to reconcile with one another experience less aggression going forward.[2]

When a young snow monkey named Imo' innovated cleaning her sweet potatoes by dipping them into the sea, the adults in her troop were slow to take notice. They lived together on a tiny islet in Japan, but only two adults in the troop copied her behavior over the next five years. The young monkeys, though, the other children and subadults, watched and learned. Five years on, nearly 80 percent of the juvenile monkeys were cleaning their sweet potatoes in the style of Imo'.[3]

During childhood, we learn how to be. We also learn who we are, and we dream about who we might become.

Humans are not blank slates, but of all organisms on Earth, we are the blankest.[4] We have the longest childhoods on Earth,[5] and we arrive in the world with more plasticity than any other species—meaning that we are the least set in stone. Software, which is the interplay of experience and knowledge with capacity, is more important in humans than in any other species. A profound demonstration of this can be seen in the peopling of the Americas. A handful of ancestors came into the New World with Stone Age technology, and diversified into hundreds of cultures across two continents, inventing writing, astronomy, architecture, and city-states along the way, the pace of change being far too rapid to be attributable to genes. It all took place on the software side.

Our ability to learn language is part of our hardware. Nearly all human babies have this competence latent in them. Which language a baby will speak, however, is entirely context dependent: that's software. Furthermore, we quickly lose some of our ability to hear and construct the phonemes and tones of languages that are not in our environment, regardless of our particular ethnicity or lineage. Just as we are born with more neuronal potential than we use (most of our neurons die off before we become adults), we are also born with more linguistic potential than we use, and some of it is lost during childhood. We are born with broad potential, and that potential narrows over time.[6]

On its surface, the pairing down of initial capacity may seem like a tremendous waste. So why do we do it? The answer is that when we are born we are in a mode of exploration. We cannot predict ahead of time precisely what neurons we will need, or what language we will speak, so we are born with a surplus of capacity. This permits us to optimize our minds to whatever world we find ourselves born into, without the need for prior knowledge. We are born to explore the world around us, discover its secrets, and structure our minds accordingly. Once this job is done, we shed our surplus capacity, lest it become a metabolic liability, all cost and no reward.

Humans are social, with long life spans, and have overlap between generations: grandparents, parents, and children may all live in the same place at the same time. These characteristics also apply to the other apes, to the toothed whales (dolphins and orcas) and to elephants, to parrots and corvids (crows and jays), wolves and lions, and more.

All species that are social, are long-lived, and have generational overlap also tend to have long childhoods. These other species' childhoods come with tantrums and play, emotional depth and cognitive capacity, just like ours. The adults that develop from those children of other species have social complexity that is recognizable to us humans as well: spinner dolphins choreograph elaborate group hunts;[7] New Caledonian crows share information among their friends;[8] elephants grieve.[9]

Spending time as children allows animals to learn about their environment. Therefore, stealing childhood from the young—by organizing and scheduling their play for them, by keeping them from risk and exploration, by controlling and sedating them with screens and algorithms and legal drugs—practically guarantees that they will arrive at the age of adulthood without being capable of actually being adults. All of these actions—almost always well intentioned—prevent the human software from refining our crude and rudimentary hardware.

Absent childhood, animals must rely more fully on hardware, and therefore be less flexible. Among migratory bird species, those that are born knowing how, when, and where to migrate—those that are migrating entirely with instructions they were born with—sometimes have wildly inefficient migration routes. These birds, born knowing how to migrate, don't adapt easily. So when lakes dry up, forest becomes farmland, or climate change pushes breeding grounds farther north, those birds that are born knowing how to migrate keep flying by the old rules and maps. By comparison, birds with the longest childhoods, and those that migrate with their parents, tend to have the most efficient migration routes.[10] Childhood facilitates the passing on of cultural information, and culture can evolve faster than genes. Childhood gives us flexibility in a changing world.[11]

Learning to Navigate Backflips and Traffic

The human desire to do a backflip must be almost as old as bipedalism. The ability to record oneself learning to do backflips is rather more modern. One can find myriad YouTube videos that chronicle a young person's attempts to stick the landing on a double backflip. It takes the inclination to try, and numerous attempts over many days, weeks, maybe months. It takes the willingness to risk injury, and perseverance in the aftermath of failure. There is no guarantee as to how long it will take, and no sure path to success. If you won't accept all of these things, you are unlikely to end up being able to do backflips.

Sitting in a room listening to a talk about backflips is not going to result in you doing any backflips either, although you might learn how to answer questions about backflips. You might learn to *sound* like an expert, while in possession of no actual expertise.

Children learn through observation and experience. While various cultures, including WEIRD ones, focus increasingly on direct instruction (which becomes formalized as school), cultures from the Navajo to the Inuit avoid teaching when at all possible.[12]

Children learn from their parents, from their siblings, from their extended family and friend groups. Siblings have historically been particularly corrective forces, as they tend to be brutally honest (and sometimes just brutal) when their brother or sister does something poorly, or has an error in judgment. The ways that children enforce their own views of appropriate behavior on other children may look mean to adults, but when children are actually allowed to roam freely, in groups, and engage in long periods of unstructured play, the bullies and jerks are more likely to lose power than gain it,[13] and everyone learns how to both create and follow rules that work. Across cultures in which play has been observed, even very young children who are allowed to engage in open-ended play in potentially dangerous areas with no adult supervision tend to resolve disputes quickly among themselves, and rarely have accidents.[14]

Compare this to a modern recess at school: All play is supervised, children are often restricted from playing or creating games that limit who or how many people can play (because that would be "exclusionary"), and any disagreement between children is immediately arbitrated by an adult.[15] Children growing up restricted to environments like this will not be capable adults.

In Quito, a bustling metropolis of fast-moving vehicles and inconsistent adherence to traffic laws, we watched a tiny child, maybe four years old, navigate a complex intersection entirely on his own. After crossing several lanes of traffic—completely safely, without having required any traffic to stop for him—he entered a small store, purchased a bag of fruit, crossed back through the same intersection, and disappeared into an apartment building where, presumably, a mother or aunt or other adult was waiting for him, having sent him off to procure food. At the time, our own children were eleven and nine, and we didn't trust *them* to navigate that intersection on their own. They had never been exposed to such a situation before—how could they already be savvy to its risks? They were, however, already savvy about the Amazon rain forest, and we allowed them to explore the jungle in a way that the Quito child would not have been safe doing, as he had almost certainly never spent time in the Amazon.

It is a fine needle to thread, giving children enough space to make their own decisions and mistakes, and protecting them from real danger. Our societal pendulum has swung too far to one side—to protecting children against all risk and harm—such that many who come of age under this paradigm feel that everything is a threat, that they need safe spaces, that words are violence. By comparison, children with exposure to diverse experiences—physical, psychological, and intellectual—learn what is possible, and become more expansive. It is imperative that children experience discomfort in each of these realms: physical, psychological, and intellectual. Absent that, they end up full-grown but confused about what harm actually is. They end up children in the bodies of adults.

Children are perfectly designed to acquire, and to want to acquire, the skill set that they will need as adults. We moderns have disrupted this to a remarkable degree. Children will self-program, if we let them. Similarly, adults will provide a road map to the environment in which their children are growing up, unless market forces intervene (a point to which we will return at the end of the book). Buying snake oil, in the form of authoritative parenting books (well intentioned though they may be), is now considered a badge of good parenting, but it should not be (we say that knowing that we are, in part, presenting as authorities offering parenting advice). Meanwhile, trusting yourself to do right by your child, and to let your children explore and take risks, gets the side-eye.[16] This is backwards land.

Plasticity

Childhood—and by extension parenting—is an interplay of love and release, of holding someone close while also giving them freedom to explore, even perhaps to leave. In biology, we speak of *plasticity*, often *phenotypic plasticity*, to refer to the many outcomes that are possible from the same starting materials. Roughly speaking, a genotype (say, the alleles for brown eyes) produces a phenotype (the actual brown eyes). Phenotype is the observable form of an organism. For many traits, though, a particular genotype encodes information for a *range* of possible phenotypes,[17] and interactions with the molecular, cellular, gestational, and external environment determine what phenotype will actually be produced.

Phenotypic plasticity allows individuals to respond in real time to changing environments, to avoid being canalized into set patterns and lifeways by their genes.

The skulls of dominant wild hyenas are big and robust, with large sagittal crests on top and broad zygomatic arches at their cheeks. Both of these structures provide places for muscles to attach—much needed if you're in the business of asserting your dominance with your teeth. Compare this to the skulls of hyenas born and raised in captivity, which have no such structures.[18] The different environments of wild versus captive hyenas affect what form (morph) they have.

In the same vein, human children who chew soft, processed foods have smaller faces as adults than those who grow up chewing hard, tough food.[19]

Spadefoot tadpoles can grow slowly into omnivorous morphs, or if they are tightly packed and running out of time and space in the ephemeral pools in which they live, they can grow more quickly into larger, fiercer cannibal morphs, and feed on each other. Which morph a spadefoot tadpole develops into is entirely context dependent.[20]

When temperatures soar, zebra finches communicate this to their unhatched chicks. Zebra finch chicks, whose parents "told" them about high temperatures while they were still in their eggs, alter their begging behavior as nestlings, and when they become adults, they prefer hotter nest sites.[21]

Even our critically important aortic arch, the first arterial branch off the heart that takes oxygenated blood to the body, has several common anatomies within human populations, which can develop from highly similar genetic starting gates.[22]

Plasticity provides the possibility of alternate phenotypes, often through simple rules that do not prescribe precise outcomes. The result, ever more so with increasing levels of complexity, is exploration of new territory, literal and metaphorical.[23]

One place that plasticity manifests in humans is the wide variety of approaches to parenting across cultures. In Tajikistan, babies and toddlers are restrained for hours on end in cradles known as gahvoras. Gahvoras are treasured within families, and passed down between generations. Tajik children are the center of family life; mothers, grandmothers, aunts, and neighbors are always available, and immediately respond to cries from a cradled baby with food, singing, or other comfort. Counter to Western expectations, though, within a few weeks of babies being born, they are placed in gahvoras, provided funnels and holes through which to pee and poop, and their legs and torsos are tightly bound.[24] Children thus cradled can move their heads, but little else. These children, with little experience crawling or attempting to walk in their infancy, do not walk as early as children raised in the West. The World Health Organization's formal expectations for when children start to walk are between eight and eighteen months,[25] and yet Tajik children may not walk until two or three years of age.[26] Are the Tajik babies dullards, or physically incompetent? No.

In contrast, children in a rural Kenyan village sit and walk earlier than Western babies tend to.[27] Are the Kenyan babies inherently destined for greatness, their precocious motor skills predictive of early mastery across domains? Also no.

Variations in baby-raising culture across humans exemplify some of the great plasticity that humans have. Kenyan babies walk earlier than Western babies—but all but the most severely disabled Western babies learn to walk, soon enough.

WEIRD parents are not just focused on our children, we are focused on the metrics that are easily recorded and conveyed to others: the *when* of our child's first smile, word, or step. Once we have such metrics in hand, we are easily confused into imagining that the *when* is a critical measure not just of health, but of future capacity. Once again, the easily measured thing—the calorie, the size, the date—becomes an inaccurate stand-in for a larger analysis of the health of the system. By believing in the false notion that *when* a benchmark is met is the salient measure of health and progress, we play into our modern fear of risk. It is *risky* for my child to miss a benchmark. It is *risky* for me not to force my child to meet arbitrary deadlines. Such parental focus can instill fear in our children, which they carry forward as an aversion to risk.

Fragility and Antifragility

Humans are antifragile:[28] We grow stronger with exposure to manageable risks, with the pushing of boundaries. As we grow into adults, exposure to discomfort and uncertainty—physical, emotional, and intellectual—is necessary if we are to become our best selves.

Just after fertilization, the zygote is incredibly fragile. A large percentage of pregnancies end in early miscarriage[29]—and in many of these cases, it happens so early that the woman never even knows she was pregnant. With every passing day, the zygote becomes more robust, more resilient, and more capable—but even at birth, the baby isn't exactly ready to rock and roll. We are born unformed, in need of active, nearly constant parental care for a long time.

From extremely fragile zygote, to fairly fragile baby, to ever less fragile child and young adult, the goal for the individual, and their parents, is that they become antifragile, not merely unfragile. In part, this takes recognizing that development is a continuum. Just as mothers-to-be are advised not to feed their fetuses alcohol by drinking when pregnant, we don't give alcohol to babies or toddlers. The line grows fuzzier and fuzzier over time, though, and at some point it becomes OK for a young person to drink alcohol, because the anatomical and physiological systems are developed enough to deal with the insults and injuries that drinking brings. Similarly, we don't expose our children to physical or emotional risk in utero if we can help it. Birth seems like this bright line, a clear and unambiguous border—and it is in some regards—but the less of a bright line we can make it for the baby, the stronger and more antifragile that baby will grow up to be.

"Expose children to risk and challenge" is therefore a rule, like so many rules in complex systems, that is context dependent. So while exposing your child to ever greater risk as they grow up is integral to their becoming antifragile, you cannot do so by simply throwing them in the deep end. First, you must make sure that your child knows, at the deepest level, that they are loved, that you have their back, and that no matter what, if they are in trouble, you will do everything possible to come in and get them.

Bond tightly with your children early on. As we have shown, different cultures do it differently. We are fans of attachment parenting: carrying your child as you move through the world, so that they see what you see, and are in literal contact with you; and co-sleeping with your baby (which, contrary to some reports, makes having an infant easier on parents, not harder). When your baby cries, go to him, assure him that he is not alone. A child thus treated is likely to have the confidence, pretty early, to go out adventuring, because they know that someone—their parents—have their back, no matter what.[30]

So when some parents try to make their babies resilient by putting them alone in dark rooms and expecting them to learn to comfort themselves, those parents are not understanding what sort of being they have on their hands. There is nothing in our millions of years of evolutionary history that should lead an infant to feel secure alone in a room. The screaming that results may actually not just be maddening to parents, but may also be a way for a baby to assess whether or not she's safe from danger. If she's safe (and helpless infants, unlike college students, *do* need safe spaces), she can go about her business learning how to be a human. It may not look like she's learning much, but she is, and the neural circuits that she's laying down now will almost certainly look different if they emerge from a position of: "I'm confident and secure because I am taken care of," as opposed to "I don't know what's what." The latter is likely to produce fear and anxiety.

The fact that the child has no idea what she is doing, or why, does not make it any less real, or any less evolutionary. The calculus involved in the building of a snail's shell is real, but no sane person concludes that snails are consciously doing calculus.

The younger the child, the more secure and safe she needs to know that she is, which creates the inner strength and resilience to go out and explore sooner, and with greater skill and courage. Just because a parent knows that he adores his child, and would let harm come to himself in order to protect her, does not mean that this is known by the child. Tiny larval forms can't know that yet. The only inputs a baby can take in are: When I communicate my needs, are they met? Do I have evidence that my parents are present when I call for them?

Of course, children will quickly learn to test the system, and to try to game their parents. Parents and children are together for the long haul, and children are selected to figure out parental moves, and to attempt to manipulate them. Indeed, manipulation begins before birth.[31] A fetus is selected to extract resources from its mother, even as the mother is selected to provide for her child, while also keeping some in reserve, both for her own health and for that of future children.[32]

Static rules don't work with children. Rules have to be nimble, able to change as the child matures, and responsive to both the needs and the tactics of the child. That being said, as early as possible—and really, far earlier than there is any expectation that the child can understand what you are saying—talk to your children as if they are mature and responsible beings. Hold them responsible for their actions, and for ever more of their needs as they grow up. Give them real work to do, not busywork. Do not make false threats ("If you keep doing that, I'll turn this car around!"). Always make sure that they know they are loved.

With full understanding that luck and timing are beyond a family's control, and that even the best-laid plans and parenting are no guarantee of success, allow us to tell you how this has looked for us. With our own children, we have expected them to make their own breakfasts and lunches on school days since they were in elementary school, as well as to feed the pets daily and do their own laundry every week. We also, gradually, exposed them to a wide range of risks. By the time they were ten years old, they were trustworthy on top of mesas in Eastern Washington, with coral snakes in the Amazon, in forests and surf at various locations (but less competent in cities). When they did get hurt in minor ways, we didn't put bandages on "boo-boos"; instead, we told them to get up when they fell and get back on the bike or scooter or up the tree.

But when they were tiny, one of us was usually touching the child—wearing him, carrying him, sleeping with him. Now they are adventurous and polite, with senses of humor and justice. They know to honor good rules and question bad ones. We have told them that sometimes we'll make mistakes and give them bad rules, but we are 100 percent on their team, and they should ask why our rules are what they are, but it is counterproductive to simply break them for the sake of breaking them. For the most part, they don't.

One set of rules that many WEIRD parents have, but that are frequently broken by children, surrounds bedtime. How do you increase the chances that you will end up with children who literally never come out of their bedrooms after bedtime, as we did—thus securing hours, weeks, months of time for the other adult(s) in your life? For the first year-ish of their lives, our boys slept with us or next to us, and when they cried, we responded quickly. Sometimes it felt endless, to be sure, but in short order, they didn't cry much. Once they were sleeping in their own room, we had family nighttime rituals, like reading to them, but we also made it clear that bedtime was bedtime and that they should not game us. When it was time for bed, we tucked them in, and neither of them ever did come out of their room with nighttime demands. We believe that this is in part because they knew that we were there, and that if they really needed us, we would come.

Play and Tinkering and Sport

Humans are both competitive and collaborative. We cannot be human without both of these things, and unstructured play reveals both in children.

Play, it has been proposed, serves to enable mammalian children to develop kinematic and emotional flexibility for use in unexpected, and uncontrollable, situations.[33] In young golden lion tamarins—small Brazilian monkeys with orange manes—play can be wild and raucous. It costs metabolic energy, and forces adults to be vigilant on their behalf, as the risk of predation from hawks, big cats, and snakes is real for these diminutive monkeys.[34] Despite these costs and risks, complex play persists, and so—hearkening back to the three-part test of adaptation introduced in chapter 3—play must be adaptive.

Play can look so many ways. Broadly speaking, play can explore the physical world, the social one, or some combination of the two. There is terrific value in tinkering, in taking physical objects and moving them through their paces, in taking them apart and seeing if they go back together again. We are both old enough to remember hobby shops and Radio-Shack being places that facilitated such investigation; the decline and demise, respectively, of such spaces, in combination with widespread replacement of mechanical parts with electronic ones (in everything from cars to toasters), means that this kind of play is more difficult to come by in the 21st century. It is well worth seeking, however. This investigation of mechanical space is no less exploratory than a hike off-trail through physical space. Many girls are more likely to want to explore explicitly social space—staging tea parties, in which they act out the words and intentions of their guests, which may be dolls and stuffed animals, in advance of having real guests with whom they can interact—and that, too, is exploratory.

Formal sport often brings both together, especially team sports. Team sports can bring the physical and social together in a fun and creative way, and are a valuable platform for exploration. They are not for everyone, but sports are one way to ensure physical skill, and physical skill facilitates mental clarity and strength. That said, team sports are not a complete replacement for either unstructured play or physical engagement with the world that most would call "work." Work must be done, and children are well served by doing some of it. For instance, given that fences exist, they are built by someone, and it is too easy for those who have never built one to imagine that doing so is simple, or banal. In families of white-collar workers, if children engage in physical activity only when parents ferry them to specially sanctioned places and times for formal sport, the illusion is created that real physical work is always an option, never a necessity. While that may serve your class aspirations (and it may currently reflect the reality of your life), it does not serve your child. Sport is valuable, *and* it should not fully replace physical work.

So formal sport is valuable, and physical work is, too, but deeper yet is simple play, with no top-down enforcement of rules. When children play a pickup game in their neighborhood, making up the rules as they go, or modifying rules for established games for whatever court and gear they have at hand, they are learning deep truths from play. Those children learn even more if they are of many ages and skills. Younger children gain access to activities they could not do alone, are able to observe activities they are not ready for, and receive mentorship and emotional care beyond that which their peers could offer. Similarly, older children in mixed-age groups gain practice in nurturing, leading, and mentorship, and often get inspiration for creative activities.[35]

Remember Chesterton's fence—that irritating object that you should not remove until you know its purpose—and consider Chesterton's play, in all of its messy diversity. Eradicate it at your—and your child's—peril.

On the Dangers of Apparently Animate Objects That Can't Respond

Do not let inanimate objects babysit your children. Leave a child alone with displays that look and sound alive—be they human actors or animations—but are not alive, and so do not respond to living beings, and the child learns all of the wrong lessons. Why is there an uptick in diagnoses on the autism spectrum now?[36] We posit that it is, in part, related to the number of children who were raised staring into screens animated with creatures that seemed to be alive but weren't. Those seemingly alive creatures, which cannot and therefore will not respond to a child's looks or gestures or questions, send the message to the developing brain that the world is not an emotionally responsive place. What is a child to make of this world? How is a child to develop a nuanced theory of mind—the ability to attribute mental states to others, and to understand that others can and do have desires and opinions that are different from one's own?

Our ability to recognize that other individuals are both different from self and yet equally deserving of respect and fair treatment is not unique to humans. Baboons, for instance, show a deep theory of mind. A female baboon can accurately assess whether threatening vocalizations from another female are directed at her, based on what social interactions the two individuals have recently had. A baboon understands that, when another individual is looking at food, they are likely to defend that food should it be threatened. Yet baboons also fail at tasks that would seem obvious to humans—mothers routinely carry their babies on their stomachs, and when moms make water crossings between islands, they continue to do so, which sometimes drowns their submerged infants.[37]

Humans engage in theory of mind more often and more deeply than any other species. We interact with inanimate and animate objects differently, and learn not to ascribe intent to those that do not react. Letting inanimate objects babysit your young child risks sending the child the message that others in the world are neither responsive to nor deserving of respect and fairness.

Legal Drugs and Children

In combination with restricted access to risk and play (helicopter parenting), and using screens as babysitters, the diversity of legal drugs that are now regularly given to children helps create a perfect storm of societal factors that are damaging our children.

The considerable rise in mood-altering and behavior-modifying pharmaceuticals being given to children in the last several decades[38] is, we posit, in part a response to children resisting school culture, which we will explore further in the next chapter. Boys are more likely to get diagnosed with ADHD and prescribed speed, which "allows" them to focus, and increases the chances that they will tolerate sitting still, facing forward, in neat rows. Because rough-and-tumble play no longer suits our delicate sensibilities as a culture, we prefer to drug our children into submission. Girls, on the other hand, with less proclivity toward "acting out" and greater proclivity toward being agreeable and anxious, are more likely to get prescribed antianxiety meds and antidepressants. Most school seems better suited to girl ways of being and learning than to boy ways of being and learning,[39] but that doesn't mean it's healthy for girls, either.

The conditions that boys tend to be diagnosed with are often classified as learning disabilities or, more recently, the less fraught term *neurodiversity*. We posit two things about neurodiversity.

First, except for rare, extreme examples, many people who exhibit "neurodiversity" benefit from trade-offs that allow them enhanced insight or skills in other areas. There is also value simply in being the "rare phenotype," in looking at the world differently from how the majority sees it. This logic applies not just to people who have autism spectrum disorder, especially if they are high functioning, but also to people who have ADHD or are dyslexic, dysgraphic, color-blind, left-handed, or more.[40] Given a choice, you might not choose any of these traits for yourself or your children, but that preference may say more about our inability to understand trade-offs—especially cryptic intellectual trade-offs—than it does about what is actually beneficial for individuals, and for society.

Second, while learning differences aren't inherently good *or* bad, they can serve to break bad educational relationships. A good teacher-student relationship is liberating, but a bad one can be devastating, and this is made more likely by the quantification of teaching that can turn teachers into seal trainers rather than holistic educators. Once education has become highly canalized—directing people unswervingly into banal and generic choices—the canals themselves become toxic. Having a learning disability can free a person from even being *able* to play in the toxic canals, which can force such young people to forge their own educational path. This provides a view not only into the current metric-heavy system—one that too often fails to reveal wisdom or capacity—but also into a better, alternative future, one in which there are multiple routes to becoming successful, productive, and antifragile.

But the pharmaceutical industry has found, in neurodiversity, yet another opportunity to profit. Because having quiet and compliant students suits schools that have too many children and too few resources, much neurodiversity is now being suppressed with drugs.

In our own experiences teaching college undergrads for fifteen years, we received health histories from all of our students nearly every quarter, in advance of taking them on multiday field trips to the scablands of Eastern Washington, the San Juan islands, the Oregon coast. By the late aughts (2008, 2009), some of our academic programs had a populace in which more than half of the students were still on or had as children been on mood-altering pharmaceuticals—again, typically (but not always) speed for the boys, antianxiety and antidepression meds for the girls. That number did decline some in the years that followed (although this happened in parallel with an uptick in doctor-prescribed exogenous cross-sex hormones and hormone blockers), and there was always a substantial minority of students on something prescribed by a doctor. Many of these students were actively trying to wean themselves from these cocktails; some of them were even successful.

Does a Butterfly Remember How to Be a Caterpillar?

As a child grows, from toddler to early child, into the tween and teen years, that child changes. It is not just the child's anatomy and physiology that are changing, though—her size and shape, her proportions. Her brain is changing too, her psychology. These changes, this process by which we learn how to be adult humans, are indeed the point of childhood.

So it is particularly challenging being a child in an era when there are permanent reminders of an earlier time. When, as a thirteen-year-old, you see a photograph of yourself as a six-year-old, you know that you both are and are not the same person you were then. You are in the act of transformation. As humans, we can and do continue to transform throughout our lives, but the most intense period of transformation is childhood, just as identity is being formed. The fact of transformation can make it challenging to reconcile who you were in early childhood with who you are in late childhood. More challenging yet is reconciling who you were in late childhood, when perhaps you thought yourself already an adult, with who you are as a young adult. This is made even more difficult with a permanent record of those earlier phases always around to remind you.

If running into photographs of you in an earlier instantiation is difficult to reconcile with the new and updated you, social media make this orders of magnitude worse. If you are a middle-class fourteen-year-old in the WEIRD world today, you are likely on social media, posting evidence of your awesome self. Just a few years later, those earlier posts seem to provide evidence of what you were, even if you yourself know that the posts were, at best, curated; at worst, outright lies. Children are now competing with earlier versions of their own selves. Combine calls to "be your authentic self" with a Western cultural norm of always being right, and early social media posts are destined to confuse and thwart children as they should be metamorphosing into their adult forms.

If running into photographs that you and your peers put on social media beginning in your early teen years is difficult, it only gets worse the earlier that record starts. If you are on social media in middle school, your identity is bound to be confused and confusing. If your parents were posting images of who you were at seven, and you have those to compare to as well, it's harder still. Yes, we deserve to have photographs of our children at all stages of development. In general, those photos should not be on display for everyone to see, unless they clearly represent a particular moment in time that is not meant to be universal.

We are being solidified by modernity into states that, in prior eras, would have been more ephemeral. Consider the philosophical question, first introduced by the ancient Greeks, of the ship of Theseus: If over time this ship has a plank replaced because of rot, then another, and another, such that ultimately every single original part has been replaced, is it still the ship of Theseus? Is it, in fact, the same ship? With an individual organism, even more than for a ship, the answer might be both yes in one sense and no in another. Yes, we have a continuous lifeline from our birth until our death. Yet the transformations that occur, most intensively as we move from childhood to adulthood, mean that we are not the same beings as we were, and that if we try to hold ourselves to a previous identity, we will restrict our future.

So, does a butterfly remember how to be a caterpillar? It does not. The incomplete nature of memory, in this case, is not a flaw in the programming. There is no need for a butterfly to remember its caterpillar life. Similarly, adult humans remembering precisely how they thought about the world when they were younger is not, in general, necessary to living a good life—especially if those thoughts and images are doctored, not reflective of what was actually true. Being constantly reminded of what we looked like, how we acted, what thoughts we decided to post on social media when we were younger and different is actively getting in the way of the ability to grow up. This applies to adults as well as children.

The Corrective Lens

- **Do not expect your children to keep up with the Joneses.** Some developmental "delays" are indeed delays, and indicative of physical or neurological problems. But development is wildly plastic, and doesn't always happen in the order you expect it to, or at predetermined moments. Don't panic if your kid isn't reading in the second grade. The chances he's going to grow up illiterate are slim to none. Earlier is not necessarily better. Early walkers, talkers, or readers do not inherently become more adept, smarter, or more productive adults.
- **Encourage active engagement with the physical world.** Do this mostly by modeling it, but also by making opportunities and, to some extent, toys that make it easy and fun, available. Allow mistakes. Expect accidents, falls, minor injuries. Be prepared for the possibility of larger injuries. Remember that people do not learn exclusively from being told what others have learned—

especially physical truths. They have to have the close calls themselves.

- **Do not let inanimate objects babysit your children**, especially if those objects are masquerading as animate ones.
- **Do let children play without adult supervision** as early and often as possible. This includes in game and sport situations where there are established rules.[41]
- **Consistently follow through on your promises**, both positive and negative. Do not make a threat (e.g., If the screaming continues, I'm going to take away your toy.) and then not follow through. Better not to make such threats in the first place, but if you do make them—and we nearly all do, sometimes—make sure that you deliver.
- **Expect that static rules will get gamed.** Becoming an adult is, in part, about learning what the system is, where its weaknesses are, and how to take advantage of those weaknesses. Children learn this in the system that their natal home provides. Make honorable systems, listen to children when they have grievances, take them seriously from an early age, but do not pretend to them or yourself or anyone else that yours is a friendship rather than a parent-child relationship. Stop manipulation every time that it starts.
- **Do not helicopter or snowplow your children.** Let them make their own mistakes. At the same time, make clear rules. One that we set was this: "You are allowed to break an arm, a leg, a wrist, an ankle. But you may not break your skull or your back, or impair your senses." This allowed our children a sense of what kinds of risks were acceptable to take, and also what kind of plans B, C, D, and so on they needed to have in order to protect their brains and central nervous systems above all else.
- **Do not spoil your children;** instead, give them responsibilities early on. A child who is catered to constantly comes to expect it, and is destined to be both dissatisfied with the world beyond his natal home and unwilling and likely unable to do much for himself.
- **Let your children in on (almost) every conversation.** Reward your children's inquisitiveness with conversation, and do not dumb ideas down for them. Obviously, there are some things that are

inappropriate at various developmental stages and ages, and what you personally decide is appropriate and when will vary by person, but in general, assume that your child is smart and can handle the content of an adult conversation. Don't try to make them interested, just have it, demonstrate through your actions that this is what is valuable, and they will come to value it too (just like with food). Similarly, involve them in tasks that are actually useful, and engage them in those tasks in a way that enhances their understanding of the world.

- **Let siblings (and friends) teach each other, and do not intervene whenever they have a disagreement or altercation.** If they ramp up their arguments so that you have to get involved, do not reward such behavior. They should be resolving their own disputes as early as possible.

- **Let your children sleep.** Sleep plays a crucial role in brain development, and when synapses—the connections between neurons—are being generated at a very high rate, sleep expands in scope as well.[42]

- **Do not succumb to dominant parenting expectations.** Most of them are silly—at best unnecessary, at worst actually harmful. Listen to your own self, and don't let parental peer pressure get you doing things that you take issue with, or that feel wrong for your children. (Examples might include: constant playdates; scheduling lots of meetings and lessons.)

- **Do not make a habit of displaying your children on social media.**

- **Provide ample free time for your children** and, if possible, allow them to explore unwatched during that time (many moderns, however, live in situations that do not allow for this).

- **Be the kind of person you want them to become.** Monkey see, monkey do. Don't be surprised if your children eat processed food and ask to buy things at every store, if that's what they see you doing.

Chapter 10

School

Children across cultures and through time have managed to grow to adulthood and learn to become functioning members of their society without the necessity of schooling. Jump to the 21st century and we find a world where childhood without schooling is unthinkable.

> David Lancy, in The Anthropology of Childhood: Cherubs, Chattel, Changelings[1]

The primary goal of real education is not to deliver facts but to guide students to the truths that will allow them to take responsibility for their lives.

> John Taylor Gatto, in A Different Kind of Teacher: Solving the Crisis of American Schooling[2]

THE WESTERN AMAZON WAS EXPERIENCING A DROUGHT.

Our class of thirty undergraduates, plus our boys, then nine and eleven years old, were staying at a remote site on the Shiripuno River. The Shiripuno flows into the Cononaco, which feeds the Curaray, then into the Napo, and finally the Amazon itself.[3] It was searingly, blindingly hot. Bret, our boys, ten students, and our skilled guide Fernando were hiking through the jungle to find a salt lick, where animals congregate to replace precious nutrients. It's always dark in the understory, but the light fell even more as, after far too long without rain, water began to pour from the sky. Soon the trail was a stream, and soon after that, it had disappeared entirely. Fernando advised the rest of them to stay put while he retraced their steps and refound the trail. Winds kicked up, whipping the canopy branches wildly. As the monkeys fell silent, the forest itself began to howl, trees joined by lianas pulling against one another, the tension forcing high-pitched squeals. Into the middle of this came a distinct, sharp *crack*.

Seeing movement just before it crashed down on them, Bret dove onto the boys, covering them and flattening them to the ground. They disappeared under a massive canopy tree. Buried under foliage and branches, they had been hit by the crown, but not by the trunk. Immediately, the muffled shouts of students—all of whom were fine—reached them. *Zack! Toby!* They yelled, stricken. *Zack, Toby!*

After a few minutes, Zack, Toby, and Bret climbed out from under the tangled canopy, unscathed but for some ant bites. The winds were still high, rain streaming down, the forest floor now a maze of fast-moving streams, but everyone was safe.

Just a few weeks later, an accident in high surf in the Galápagos nearly killed Heather and the boat captain. The boat accident could easily have killed all on board, including eight of our students, some of whom were present for the tree fall at Shiripuno as well. That story is long and terrifying, and we have documented it elsewhere,[4] but some of its lessons are the same: Keep your wits about you. Believe that you can rather than that you cannot. Build deep community and, having built it, trust that it will be there for you.

We had selected the students for this study abroad program for their combination of intellectual skill and curiosity, physical and problem-solving aptitude, and community-mindedness, not for any latent parenting skills or interests. Here many of them were, though, acting almost parental. Central in our approach to education was the building of community, of actual and genuine relationship, not just between students, or between students and faculty, but also—when on this extended study abroad trip—between our school-aged children and our college students, many of whom were far closer in age to our boys than to us, and some of whom were closer to our age. This was education for everyone—for our students, for our children, and for us.

School is brand-new in our evolutionary history. It's newer than agriculture; newer than written language. Like all social, long-lived organisms with long childhoods and overlap between our generations, we need to learn how to be adults. That is different, however, from needing to be taught.

Not only is school rare in human history—so is teaching.[5] There is some evidence of teaching in species outside of humans, and the examples that we have are fascinating.

In many ant species, foragers that have discovered something worth knowing—a food source or a possible nest site—convey this to others by running in tandem with them, guiding them to the new opportunity. Knowledgeable foragers could just hoist their naive nest mates on their backs and carry them to the destination—it's faster, and indeed, sometimes they do just that. But the ant being carried will have a more difficult time learning the route this way, in part because she tends to be slung onto the back of her carrier upside down and facing backward.[6] While tandem running takes much longer for the ants-in-the-know to get to their destination, the newly taught ants end up both more informed and efficient than they otherwise would be.[7]

In species more closely related to us, meerkats hunt and eat a wide range of food, some of which, like scorpions, are both challenging to capture and potentially dangerous. Adult meerkats provision very young pups with prey that has already been killed. Over several months, adults introduce the pups to live prey, teaching them to handle and hunt prey, and retrieving any prey that manages to escape from the ever more adept pups.[8] Similarly, both cheetahs and house cats bring prey back for their young to engage with and learn from, not merely to eat immediately. Some Atlantic spotted dolphin mothers forage for longer, with exaggerated movements, when their calves are present.[9] Even many nonhuman primates—although not chimpanzees—show similar tendencies to teach their young sometimes.[10] But no other species—and no other human cultures but

WEIRD ones—has outsourced the vast majority of learning to a school environment.

In fact, many human cultures actively *avoid* teaching. Among Japanese women who dive for abalone, for instance, one woman became incensed at the suggestion that, decades earlier, her mother had taught her. She reported that her mother had shoved her away when she was just learning, told her to find her own abalone, and "practically screamed at me to move OFF and find my danged abalone BY MYSELF."[11] In cultures and situations as varied as those of the Japanese women, hunting by the Siberian Yukaghir, and the operation of power looms by 20th-century Guatemalan Maya, the skills are learned with no direct instruction. In all of these cases, teaching is not just absent, but strenuously avoided.[12]

In light of the relative rarity of teaching both in other species and among other human cultures, we should be asking ourselves: What do we need to learn in order to become our best selves? And of those things that we do need to learn, which of them need to be taught, and which can we learn in other ways—through direct experience, or through observation and practice, for instance? Put another way: What do we need school for?

You don't need school to learn to walk, and you don't need school to learn to talk.

You do need school to learn to read and write. Or rather, most people need *instruction*. Reading and writing are so new that we need an educational supplement in order to learn it. School is also useful to learn cell biology, written history, and all but the most basic math. Literacy, like math and thinking from first principles, is like an adaptive foothill, though, and once you're literate (or numerate, or adept with logic), you can teach yourself many things without requiring further school.

We can also use school to discuss texts with real people, to gain exposure to ways of thinking about and representing the world about which we were previously ignorant, and to gain experience in proposing and running scientific experiments. School is not necessary to engage in any of those pursuits, but it can be useful.

In school we also might learn what it sounds like when irreconcilable positions meet one another. This allows an insightful person to go on to do the same thing within themselves: hold two irreconcilable positions in their head at once. The value in this is immeasurable; it allows a person to learn argument by arguing with themselves, which facilitates their ability to both uncover and recognize truth. Humans are perhaps unique in the degree to which our theory of mind—the ability to understand that other living beings have points of view, and that those perspectives might be different from our own—enables us to explore contradiction and paradox. Again: paradoxes are what we see when we stand in the wrong place, processing what we see with faulty models—why would the Malagasy feast on a regular basis when they are so close to starvation? Paradoxes are the X on an analytical treasure map, inviting us to *dig here*. While the West has tended to avoid paradoxes and to find them troublesome, Eastern traditions are more likely to have embraced inconsistency. We argue that Buddhism being littered with contradictions[13] is adaptive, serving exactly the educational purpose we are advocating for. Similarly, classrooms ought to be littered with paradoxes, left in various states of interpretation, for children and older students to discover and poke at and understand.

We can also use school to hone our memories, but again, school is not required to do so. Great Argentine writer Jorge Luis Borges wrote a cautionary parable about having a prodigious memory. In it, the protagonist, Funes, is doomed to remember everything that he has experienced: "Without effort, he had learned English, French, Portuguese, Latin. I suspect, nevertheless, that he was not very capable of thought. To think is to forget a difference, to generalize, to abstract. In the overly replete world of Funes there were nothing but details, almost contiguous details."[14] Funes, in short, was stuck in the trees, unable to see the forest.

Because memory and recall are easy to assess and measure, they can easily become *the* metric that is being chased, by students and teachers and schools alike. Far harder to teach and to quantify—and at least as valuable, if not more so—are critical thinking, logic, and creativity. Memory exercises tend to drill down on detail, on facts that are unchanged by context. Trade-offs being ubiquitous, a focus on memorized details, then, will likely come at the expense of a focus on the big picture.

School can also be useful in teaching both science and art, a task made easier if the assumption is that children have latent scientific and artistic tendencies already. While people don't intuit the formalization of the scientific method, children are inclined to observe pattern, to postulate reasons for the pattern, and to try to figure out if they're right. All people are inclined to be verificationists, to look for verifying evidence of their own correctness, rather than to look for falsifying evidence that, if it doesn't show up, makes their precious idea look more and more likely. School—or an engaged parent or friend, or direct and repeated experience—can teach the value of falsification. Would that it did so more often.

Similarly, an individual cannot intuit the method by which pigments are generated for the palette, or the history of artistic movements, but individuals are inclined to observe and represent the world, in ways ranging from realistic to wholly fanciful, and do not require formal education to do so.

Left to their own devices, people reveal that they are inclined to be both scientists and artists.

What Is School?

For children, school can be understood as love and parenting commodified. Put another way: part of what school is, is parenting that has been outsourced. We have already seen many of the harms and risks of reductionism. Yet one more is that reductionism facilitates the commodification of easily quantifiable things, while tending to ignore those things that are less quantifiable. Thus, school becomes about metrics—how much, how fast, how well: did the little one read, do their multiplication tables, memorize a poem? It should go without saying that there is clear and enduring value in reading, multiplication, and poetry. A focus on speed and quantity is an error, however. What myriad things are not being learned in school, because they succumb less easily to reductionist assessments? School is based on an economic efficiency, while being unimaginative about what could be accomplished. The economics—not to mention the perverse incentives behind compulsory schooling—of school tend to fill children's heads with knowledge, without showing them a path to wisdom.[15]

Perhaps school should serve the purpose of helping young people grapple with the question: *Who am I, and what am I going to do about it?*[16] Another way of phrasing this might be: *What's the biggest and most important problem I can solve with my gifts and skills?* Or: *How do I find my consciousness, my truest self?* Done well, then, school can provide a great platform for formalizing and delivering rites of passage. Rather than focusing on any version of these questions, though, modern schooling, especially the compulsory sort widespread across the WEIRD world, is more apt to teach quiescence and conformity.

What if we took, as one of the goals of school, teaching children how to understand and hack their own incentive structures? Knock them off the low adaptive peaks that they are surely on ("I'm not good at math, languages, sports . . ." or, conversely, "I am so good at math, languages, sports, that other things fail to grab my attention.") into valleys that are uncomfortable, but from where there are many possible peaks to climb.[17]

Or perhaps school should reveal to children that fringe positions should be explored and considered, not thrown out immediately on the basis that they are unpopular. Betting against the fringe is an easy bet, usually a safe one, and when done in a tone of paternalistic indulgence, say, or authoritarian disdain, it usually shuts down dissent. While most fringe ideas are in fact wrong, it is exactly from the fringe that progress is made. This is where the paradigm shifts happen.[18] This is where innovation and creativity occur, and yes, most of it is wrong or useless, but the most important ideas on which we now base our understanding of the world and our society came from the fringe: The Sun is the center of the solar system; species adapt to their environments over time; humans can create technology that allows us to communicate across time and space, to fly, to create and explore virtual worlds. These were all impossible ideas. Laughable at the time. Those who quickly join in laughing at all fringe ideas now would have been laughing at all of those ideas in their time.

School should be fun, but it should not be gameable. A child shouldn't be able to "win" at school (although many do, and many more lose at it). Social rules and mores are learned at school, but at its base, school should be about discovering truth, both universal and local.

School is, for better and for worse, a stand-in for parents, for kin group, for those with whom the child has shared fate.

School should not, therefore, teach through fear. Risk and challenge help children learn. As with parenting, this requires early tight bonding, during which a secure base is established, which provides children the confidence to go out adventuring fairly early, because they know that someone has their back, no matter what. School that operates by fear will teach the opposite lesson.

Fear is an easy mechanism of control, and so it should not be surprising that teachers use fear to control students of all ages. As corporal punishment in the classroom fell out of favor in many (but not all) places, psychological and emotional control replaced it. It leaves fewer marks. Children are threatened with poor grades, poor test scores, and having their parents informed that they have behaved badly (which most children will hear as "you are a bad person"). The rise of metrics within a system—which are often overly simple, wrongheaded, and only pseudo-quantitative—tends to accompany a decay in social trust.[19] How can good teachers, stuck in a system of escalating metrics imposed from the outside, counteract the prevailing cultural forces? One approach, which will be more effective with older children and young adults, is for teachers to explicitly hand away their own authority by telling students not to trust them just because they are the figure in the front of the room. When a teacher then does earn the respect and the trust of her students, such that she becomes a legitimate authority figure, one with authority that was earned rather than assumed, her authority will better serve both the students and their education.

Using fear to keep children seated in neat and tidy rows, to keep their eyes forward and their mouths closed, to keep them from moving their bodies at all but for a few scheduled moments in each day—this will help create adults who are unable to regulate their own bodies and senses, unable to trust in their own ability to make decisions, and likely to demand similarly controlled environments in their adult lives—trigger warnings, safe spaces, and the like.

For young schoolchildren, one solution would be having a garden at school, and spending time in it in all sorts of weather. Frequent field trips to natural areas, and spending time actually outside rather than in the climate-controlled protection of the "nature center," help, too. Will it always be comfortable? No. Will some children be ill prepared for rain or wind or sun? Yes. Will they learn from small, early mistakes to start taking responsibility for their own bodies and fates and so get better at navigating the world? Yes. Yes, they will.

Humans are antifragile; exposure to discomfort and uncertainty—physical, emotional, and intellectual—is necessary. Preparing students to understand risk encourages them to expand their worldviews, and embrace experiences that lead to maturity. This does, however, come at a cost: understanding risk cannot completely protect individuals from danger.

In short, risk is risky! Tragedies will happen, and that's no minor thing. For those of us who have been lucky to avoid it, it's nearly impossible to imagine how a person continues on if their child has died, or if they have been involved in someone else's child dying. Tragedies that happen because someone introduced risk into a school trip are easy to point to. The story is often easy to tell, and compelling to hear. By contrast, population-level tragedies, those that happen because whole swaths of the population have difficulty navigating risk and so avoid it at all costs—this is also a tragedy, and much further reaching.

Modern school tends to protect against individual tragedies, while facilitating the larger, societal ones. Arrange all the little boys and girls neatly in rows, assign them seats, and tell them never to speak unless they are called on first, because that will make it easier to keep track of them. At the same time, at home, teach the little boys and girls that they are each the center of the entire universe, and that they may and in fact should interrupt adults at any moment, for any reason. Teach the children that temper tantrums are acceptable by caving to them whenever they erupt, and also tell the children that they are the most precious and infallible beings in existence, and as such, any criticism is a crime against their core selves.

We should not be surprised when children raised this way can make no sense of the confused and confusing messages coming at them from home and from school. Nor should it surprise us when they gravitate to the systems that are most gameable:

Mom doesn't like it when I scream or whine, but if I persist at it, she gives in as a way to make me stop? Noted.

Teacher leaves me alone if I occasionally contribute a comment in class and earn good grades, even though I'm learning nothing by regurgitating from a textbook? Got it.

Congratulations, society, you have successfully produced self-satisfied whiners who are accustomed to getting what they want, who are good at school but not at thinking, and who are, in fact, neither smart nor wise.

The World Is Not about You

Children have been harmed, through no fault of their own, by a perfect storm of societal factors that emerged in the late 20th and early 21st century, which we have already reviewed. The rise of pharmaceuticals being prescribed to children, helicopter and snowplow parenting, and the near ubiquity of screens (never mind what is on them) have all made school an even more difficult place than it once was. In the United States, add to these the economic and political forces that have reduced school funding while increasing testing, thus cutting the creativity and freedom of teachers off at the knees.

When Heather primed her students for study abroad trips—to Panama or Ecuador—in advance of embarking with them, she was trying to build not just the academic skills required for the work, but also the social and psychological ones required for extended trips outside of anything most of them had previously experienced. She would ask them: "What is your relationship with risk? And what is your relationship with comfort? Just because you can say in advance that you will be OK with the bugs and the mud and the not having access to the internet doesn't mean that you actually will be. Perhaps most important of all, though: We are going to leave ourselves open to serendipity. We cannot know what all will happen on this trip. We're going to go, and some interesting things will happen."

Those conversations included discussions of how risk is different in landscapes that haven't been rendered safe by liability lawsuits and in which medical help is far, far away. We talked about the hidden hazards of the jungle—rising water, tree falls—compared with the familiar ones, like snakes and big cats, that people are primed to be scared of.

Risk and potential go hand in hand. We need to let children, including college students, risk getting hurt. Protection from pain guarantees weakness, fragility, and greater suffering in the future. The discomfort may be physical, emotional, or intellectual—My ankle! My feelings! My worldview!—and all need to be experienced to learn and grow.

The students whom we took on study abroad trips were carefully chosen, mature, capable, smart, and adept. Even so, the inability to control our surroundings, the intentionality of succumbing to serendipity when in the jungle, threw many of them into states of confusion, which sometimes manifested as anger. Many of them believed themselves to be excited about exploration, about discovery . . . but only when it looked and felt as they had already imagined. By inculcating in children the sense that order is always better than chaos, and that being easily counted and prioritizing doing things that are easily counted is the honorable way to go through school (and therefore life, many would extrapolate), society creates adults who bristle at the unexpected and the new. Not only doesn't the jungle look or feel as you are led to believe from even the best nature documentaries, but the people on the streets of Panama City or Quito aren't what you think, the cloud forest and the people who called it home long before the Inca or the Spaniards ever arrived will surprise you, too, as will everything else, if you take your blinders off and let yourself experience the world without taking everything so personally. Mostly, the world is not about you. But you can learn from it. An education should allow you to do just that.

Higher Ed

Imagine a scholar. What do you see in your mind's eye? Put aside the phenotypic stereotypes—the glasses, the suede patches on the elbows—and realize that, in all likelihood, you imagined someone *consuming* something that had already been produced. Your iconic scholar was reading a book, in all likelihood, perhaps perusing the stacks of a library. When they enter college, students have already learned this trope. First you read, then you respond. Perhaps, someday, you too will write such a tome, which others will, in turn, sit and read, and respond to. And so the cycle continues.[20]

This model of academic activity, of what it is to have a life of the mind, to be a critical, engaged citizen of the world, has never been quite right for some academic pursuits. Science and art, in particular, often mistakenly described as at opposite ends of some imagined spectrum in the pursuit of truth and meaning, do not make their primary impact in the world through careful, thoughtful assessment and critique of what has come before. Yes, we stand on the shoulders of giants, and yes, the history of ideas and of creations that came before us is integral to what we know and think and do, but that does not mean that it ought be our primary focus, or that it is our mission.

There *are* new things under the sun, but it is the fate of every generation to think that it arrived too late, that everything is understood, and that the best response is to fall into nihilist disarray.

At its best, a college education has the potential to open up worlds—of wonder, creativity, discovery, expression, connection. For fifteen years at The Evergreen State College, a small, public liberal arts college in the Pacific Northwest, we did just that. There, the ability to go deep into complex topics, with students whom we came to know well, in classrooms and labs and the field, both near to campus and decidedly remote, provided a window into what is possible in higher ed.

Drew Schneidler, a former student of ours, and an extraordinary intellect who had had a terrible time in school rather like Bret's (and now also a friend, *and* our research assistant on this book), said to us as we were writing this: "Walking into your classroom was like walking into an ancestral mode for which I was primed, but didn't even know existed."

This statement, like nearly everything in this book, deserves a book of its own. Here are just a few of the things we learned and innovated during our time as faculty in higher education.

Tools Are More Valuable Than Facts

One of our messages to our students was this: There are intellectual tools that are more valuable than facts, in part because they are harder won. You can wield these tools with both power and precision, and with them, you may discover things nobody has even yet framed a question for.

But how do you teach tools in a vacuum? How do you actually teach people *how* to think, not *what* to think? It's easy to say, but how is it done? A well-intentioned critic might argue that the students need *some* stuff to think about, don't they? Certainly, having stuff to discuss makes things easier, but once the stuff is introduced, it is too easy for everyone, students and faculty alike, to fall into the easier, historical roles of informer and informee, the most salient representation of which is the hand raised after what seemed an inspiring discussion—*Will this be on the test?*

One piece of the puzzle is to break the carrot-and-stick paradigm. Explicitly tell students—and make sure that it is true—that they are not in competition with one another. Our students actually learned more when they collaborated with one another. There was never a "curve" looming that guaranteed that some would fail.

Another piece of the puzzle is to break the "this is the time of day that we are educated" paradigm, by leaving the classroom and spending more time together. When students and faculty do this, and break bread together day after day for several days or weeks or even months, it becomes clear that, actually, good questions show up at all hours of the day, all days of the week, and if you are traveling with an intellectual tool kit that you have cultivated through logic, creativity, and practice, you can engage such questions whenever and wherever they arise, not just in the classroom when the authority with the appropriate degree is standing in front of you, paid to answer your questions.

Intellectual Self-Reliance

> When I go outside at night and look up at the stars, the feeling that
> I get is not comfort. The feeling that I get is a kind of delicious dis-
> comfort at knowing that there is so much out there that I do not
> understand and the joy in recognizing that there is enormous mys-
> tery, which is not a comfortable thing. This, I think, is the principal
> gift of education.
>
> **Teller, in "Teaching: Just Like Performing Magic"[21]**

Imagine a professor setting out to destabilize students' preconceptions, to
make them uncomfortable with what they think they know, and to force
confrontations with self, with perception, with authority. When people are
too comfortable with what they know, and the world does not look as they
have been led to expect, they are at considerable risk—of being gamed, of
getting angry, of becoming incoherent.

Insight and growth do not happen when you are comfortable with
what you know. You can add knowledge to your foundation, like bricks in
the wall of a house you are building, and when you are done, your house
will look pretty much like what the foundation implied. For most of us,
though, that foundation that we arrived on the cusp of adulthood with is
not necessarily the base of the intellectual house that we want to live in.

Those bricks in the wall—they kill off creativity. They kill off curiosity.
Their existence makes it seem as though starting from scratch, perhaps
with no blueprints or foundation at all, is impossible. They keep us comfortable, those bricks. It's easy to keep piling bricks up, higher and higher.

The brick-in-the-wall model creates minds that are all alike, minds that
are ever less capable of generating or considering strange new ideas, minds
that are outraged by confusion, and by uncertainty.

Nearly every student whom we taught was, in the end, game to be challenged, actually challenged—told when they were wrong, told when we were
wrong, and told that they needed to learn to pose real questions and then sit
in the not-knowing for long enough to figure out how one might figure it out.

As faculty, we should aim to take our students away from the class-room. Better yet if it is somewhere with no internet, and no library—the scablands of Eastern Washington, Kuna Yala in Panama, the Ecuadoran Amazon, for instance. Once there, questions can be posed—How did these rocks get here? How do the local people catch fish? What are those parrots doing?—which are answerable, but the students will need to learn to use logic, first principles, and rigor to do so. The conversation is forced into the here and now: What answers can they generate, with their own brains rather than with the collective brain of the internet, that fit with what they observe? If they reinvent the wheel while they're at it, so be it. They will have honed their skills in scientific hypothesis and prediction, experimental design, logic. Once students can do that, not only are they becoming educated; they are, increasingly, educable.

In a successful classroom discussion, when a question of fact emerges and nobody in the room appears to have the answer already in their head, why shouldn't somebody just look it up? What harm could possibly come from establishing whether Mendeleev's first periodic table looked like it does now, how many people died in the bombing of Dresden, or when we think the first peoples in Beringia came into the New World? What harm can come from looking up answers to straightforward questions? The harm is that it trains us all to be less self-reliant, less able to make connections in our own brains, and less willing to search for relevant things that we do know, and then try to apply those things to systems we know less about.

If answering "how" questions quickly, with a few keystrokes, impedes the development of self-reliance, what of the desire to pursue "why" questions this way? It is even more likely to kill logical and creative thought. Why do birds migrate? Why are there more species closer to the equator? Why does the landscape look this way? Before you look it up—think on it. Walk on it. Sleep on it. Talk about it. Share your ideas with your friends, and when they disagree, engage the disagreement. Sometimes "agreeing to disagree" is the only route. Usually, though, something more can be learned if you dig a little. And you—and your friends—will end up more capable of understanding the world.

Calm Down, Level Up

When teaching one study abroad program in the Amazon, Heather watched rumors of the dangerous, wild, and wicked Amazon grow before her eyes. There was one other class at the remote field station where they were based, and the other professor was telling her own students about the deadly dangers of spiders, peccaries, and toads. All of the rumors were literally false, but being presented as if they were true. One in particular was about a toad that shoots toxins into people's eyes (true), causing the person thus afflicted to go permanently blind (not true). After the rumors began, one of Heather's students was squirted in the eye with the toxin of the toad in question. Panicking far beyond what she would have had she never heard the rumor, the student asked Ramiro, an excellent naturalist guide, what would happen to her. He, like all good guides, is careful, and so told her that "some people say that" this toad toxin can make a person blind. The student was fine, of course, but she went through unnecessary panic because someone was using fear and hyperbole as a tool of authority.

In the past, it was difficult to find yourself in a habitat without having an intimate understanding of it. Either you had received wisdom about it from your elders, or you had come to understand it by entering it from the edge, gradually immersing yourself in it. We moderns, though, live in such a rapidly and unpredictably changing habitat that none of us can claim to be fully native in it. We also have a problem of abrupt boundaries that our ancestors did not, a strangely clear line demarcating safety from not: the swimming pool; the garbage disposal; the curb.

Fear, anger, and hyperbole sell products, attract an audience, and are a useful tool of control. They are not, however, representative of the best that we can do as humans. Terror-inducing stories may be a hack to prompt appropriate behavior in modernity. Being able to sleep one night in bustling, cosmopolitan Quito, and the next night deep in the Amazon, is a luxury of modernity, but it has the costs of landing people in an environment for which they may have no history, and no preparation. Furthermore, the people who have landed for the first time in the Amazon generally come from the Land of Lawyers, where everything has been vetted and made safe—at least in the short term. It is a failure of education to scare people into acceptable behavior. If the ultimate goal of education is to produce capable, curious, compassionate adults, helping students stay calm and capable of reason, rather than in a constant state of alarm, is a far better route to that end.

Observation and Nature

One set of goals for higher ed ought to be to teach students how to hone their intuitions, become experienced enough in the world to reliably recognize pattern, return to first principles when trying to explain observed phenomena, and reject authority-based explanations.

This takes time together, to build relationship. Extended time—as on field trips—is a particular luxury, which not all faculty have, but perhaps all should. It takes being willing to say to students who may have been told all their lives that everything they do is commendable, "No, that's wrong. Here's why." It takes being willing to fix your own errors. Modeling for students the actual process by which ideas emerge, and are refined and tested, then rejected or accepted, allows them to move away from the linear models of knowledge acquisition that most of their schooling, and nearly every textbook, have inculcated in them.

Over the course of several trips, both domestic and abroad, we saw students rise to challenges in ways that they simply could not have done at home. We purposefully sought out field sites that were remote not just because nature is more interesting and intact in such places—more lianas climbing their way up to the light, more vine snakes mimicking those same lianas—but also because encountering nature in its least disturbed state often comes at the "cost" of having no connection to the outside world. Far from the virtual eyes that document our every move, people are revealed, to themselves and to others.

But there are risks. Ants bite. Fungus invades. Trees fall. Boats flip. Why take such risks? Is studying the politics of land use, the cultures of early Americans, or territoriality in butterflies worth it?

In the field, we watched some students descend into their own darkness, depression gripping them, and we watched as they emerged from it, stronger and more grounded. Romantic ideas of the jungle disappear with the reality of constant sweat and biting insects, and the realization that in order to see charismatic animals do interesting things, you have to get out there and fade into the forest, and then wait patiently for it to come back alive around you.

Some people hate it. They cannot abide the lack of control, the discovery that nature is not a nature documentary. Most, though, find hidden strength and unanticipated freedom.

One evening in the Amazon, our students were attempting to give research presentations under a corrugated metal roof when a squall came up. The rain pounded the roof so noisily that we had to reschedule—there was no hearing human voice under these circumstances, and no place else to go. We dispersed, some taking the opportunity to catch up on sleep, some wandering off into the forest to explore the warm, wet embrace of a tropical jungle at night during a rainstorm. If education is, in part, preparation for an unpredictable and shifting world, teaching courage and curiosity ought to be a priority.

We did also read in our classes—the primary scientific literature, books of many types, essays, fiction—and some of what we read contradicted other things that we read. Building a tool kit, though, educating minds to assess the world actively and with confidence when new ideas or data arrived, these things required getting away from texts. We went outside and engaged with the physical world, and its myriad evolved inhabitants. Louis Agassiz, one of the 19th century's preeminent naturalists, urged people to "go to nature, take the facts into your own hands, and see for yourself." By creating opportunity to go into nature—regardless of what your discipline is, and what you are trying to teach—you allow students to begin trusting themselves, rather than taking other people's words for what is true.

When you teach a small number of students intensively for two or three quarters at a stretch, as we did, education becomes personal. We told students things they did not expect:

- We need metaphor to understand complex systems.
- You are not here as consumers, and we are not selling anything.
- Reality is not democratic.

And we did not accept generic responses from them in return. We poked and prodded them, intellectually. They were forced to stretch, because repeating material back to us wasn't going to cut it, and we wanted to know something about every single one of them, so that we, too, could learn from them.

But many professors train students to be mindless workers. Another professor once told Heather, without irony, that he saw it as his job to teach the students to be cogs, because, after all, that was their fate. Faculty should know better, but with students it's different. Seduction and education are etymological sisters. Students may think that they want to be seduced, led *astray* by false praise, as it feels good in the moment. Most whom we met, though, wanted to be educated, led *forth* from narrow, faith-based belief into intellectual self-sufficiency, where they could assess the world and the claims in it from first principles, with respect and compassion for all.

The Corrective Lens

School—and, obviously, parents—should teach children:

- **Respect, not fear.**
- **To honor good rules and question bad ones.** All people run into bad rules—whether in the legal system, at home, at school, or elsewhere. If you're a parent, strive to show your children that you are 100 percent on their team—no matter the trouble they've bumped up against. Children should be free to ask why the parents' rules are what they are, but also know that it is counterproductive to break the rules simply for the sake of breaking them.
- **To get out of their comfort zone** and explore new ideas.[22] You will likely learn the least in exactly the areas where you are most certain of what you already know, whether or not what you (think you) know is actually accurate.

- **The value of knowing something real about the physical world.** When you have a sense of physical reality, you are less likely to be gameable by the social sphere. Never accept conclusions on the basis of authority; if you find that what you are being taught does not match your experience of the world, do not acquiesce. Pursue the inconsistencies.
- **What complex systems actually look like**, even if the messiness of those systems is beyond the scope of the lesson. Nature is an example of such a system. Nature provides, among other things, a corrective to the ideas that emotional pain is equivalent to physical pain, and that life is or can be made perfectly safe. Exposure to complexity is key.

Higher education, in particular, should recognize that:

- **Civilization needs citizens capable of openness and inquiry;** these should therefore be the hallmarks of higher ed. The need for nimble thinking, creativity in both the posing of questions and the search for their solutions, an ability to return to first principles rather than rely on mnemonics and received wisdom—these are ever more important as we move forward in the 21st century.[23] A misunderstanding of how work will look in the future is driving people to specialize earlier and more narrowly. Higher ed is the natural place to counteract that trend and push toward greater breadth, nuance, and integration. Students of traditional college age today cannot accurately predict what their career will look like by the time they are seventy, fifty, thirty. College is where breadth should be inculcated.
- **A university cannot simultaneously maximize the pursuit of truth, and the pursuit of social justice,** as Jonathan Haidt has famously noted.[24] This is a basic trade-off, and unavoidable. It becomes important, then, to ask what the purpose of a university is. Is it necessary that we focus on the pursuit of truth? Yes, in fact it is.
- **Social risks—intellectual, psychological, emotional—must be taken,** but doing so in front of strangers is particularly difficult. Both small class sizes and extended time together building community are correctives to anonymity.

- **Authority is not to be used as a bludgeon to shut down the exchange of ideas.** Bob Trivers, evolutionary biologist par excellence, and our mentor in college, once advised us to seek positions in which we taught undergraduates. His reasoning was this: Undergrads do not yet know the field, and so are likely to ask questions that you aren't expecting, "dumb" questions, or ones imagined to already be settled. When the educator is confronted with such questions, one of three things is likely to be true:

    1. Sometimes the field is right, and the answer is simple. Full stop.
    2. Sometimes the field is right, but the answer is complex, nuanced, or subtle. Figuring out, or remembering, how to explain that complexity or subtlety is worth the time of any thinker who deserves the title.
    3. Sometimes the field is wrong, and the answer is *not* understood, but it takes a naive view of the matter to ask the question.[25]

- **Classrooms are effectively sterile boxes removed from the world.** It is difficult to learn in such a situation, because you won't run into the things that you need to learn but that cannot be taught—things like how to survive tree falls, boat accidents, and (as we will see in the next chapter) earthquakes.

Chapter 11

Becoming Adults

BABIES ARE BORN. CHILDREN BECOME ADULTS. ADULTS GET MARRIED and have children of their own. People die. Such changes in status are marked, in many cultures, by rites of passage. Rites of passage that mark the beginning of adulthood include the vision quests of young Nez Perce men,[1] and the ceremonial cleansing, running, and dressing of young Navajo women.[2] These are symbolically important, and help young people step into their new roles. Among WEIRD people, similar moments may include your eighteenth birthday, graduating from high school or college, getting your first job, buying a house. They demarcate before and after, become a line in the temporal sand. We use ritual to make discrete the boundaries within complex systems, which are rarely so stark.

Rites of passage are useful as markers of transition—*now you are a man*, or *today you become a woman*. But they are rather uncommon among WEIRD people, less ritualistic, and this has contributed to our losing track of the characteristics of adulthood. Historically, adults were those who knew how to feed and shelter themselves, how to be constructive and productive members of a group, how to think critically. This knowledge does not magically accrue with age, though. It must be earned.

Recall the three-part test of adaptation that we laid out in chapter 3, in which we suggested that if a trait is complex, has energetic or material costs, and persists over evolutionary time, it is an adaptation. Focusing on the last element, the element of time, and putting it in terms of cultural evolution: if a trait has persistence over cultural time, it is likely a cultural adaptation. That does not, of course, mean that is inherently good, for individuals or for society, or that the conditions that rendered it adaptive in the past haven't changed, rendering it neutral or maladaptive now. In general, though, if we use caution when changing the old—invoking, perhaps, Chesterton's traditions—we are less likely to dismantle something that turns out to have been doing important work for us, and for our world.

Across cultures, rites of passage therefore provide clear signals to you, the individual, about how far along you are and what society can expect from you. Without these markers, we are more likely to end up with widespread confusion—thirty-year-olds who are effectively children, unfamiliar with responsibility, and eight-year-olds who are granted adultlike status with regard to their ability to determine, for instance, what sex they truly are. Rites of passage thus coordinate society with respect to what is expected of individuals at various stages of development, and they exist in two forms: temporal (age) and, loosely, merit (earned). Age is a rough guide to what a person should be able to do, and merit is a specific guide for what an individual is capable of or, in the case of marriage, signing up for. These have been abandoned or corrupted across WEIRD culture. Temporal rites are loosely and inconsistently applied, and merit rites are largely gameable.

People who are deserving of being called "adult" can observe themselves carefully and skeptically, and regularly ask themselves questions like these: Am I taking responsibility for my own actions? Am I being closed-minded? Am I entrenched in a worldview, and if so, why? Am I coming to conclusions independently, or have I accepted an ideology that I allow to do my thinking for me? Do I avoid collaboration that would be valuable, if it would also be challenging? Am I letting emotions make decisions for me, especially hot, intense emotions? Am I ceding my adult responsibilities, and do I make excuses when I do?

These questions all ask, in different ways: Am I doing as well as I should or could be doing? Answers will often be easier to find through one of the two categories of rites of passage. Age rites tell people what to expect of others and allow society to hold individuals accountable when they don't rise to the occasion. This teaches us to ask: Am I doing my job? Because others will be expecting that we are.

Merit rites teach us to think for ourselves and, when accomplished, view ourselves as people of knowledge and skill. They also convey this to society. This raises the bar of what is expected and what it means to "do your job." The interplay of expectation and accountability would naturally result in more self-examination to ensure you are living up to what is expected.

While it is true that we have lost track of the characteristics of adulthood, it is also true that the hyper-novelty of our world, specifically the reach of economic markets, is making it more difficult to be an adult. The market is full of con artists who want you to ignore your adult responsibilities. One of those adult responsibilities is to not spend money on every latest thing. Selling delayed gratification is rarely a successful business strategy, so it is hard to find in the marketplace. Instead, junk everything is available—junk food, entertainment, sex, news. The aggregate of the market is therefore selling infantile values, which make you a desirable consumer but a poor adult.

Absent the hyper-novelty and unconstrained market forces of 21st-century WEIRD societies, childhood is when you take in information from your ancestors, and discover the world that you inhabit, both physically and cognitively. Adulthood, then, is the phase in which you operationalize what you have learned, and become productive.

The primary modality of advertisers is to create dissatisfaction, along with the impression that others are more satisfied. To take just one well-documented example: shortly after television was introduced to the island of Fiji, teenage girls became focused on the Western ideals of beauty conveyed there, contra their own cultural norms.[3] Far beyond the shores of Fiji, the algorithms of social media have now moved into this niche as well. The ability of advertisers to create dissatisfaction is facilitated by the fact that our natural human obsession with narratives is being addressed by a narrative-generating mechanism in which the stories have not stood the test of time. Many of the narratives we hear are tailored to sell product, and are therefore what advertisers and algorithms want us to believe, rather than what we need to know.

Our narratives are also no longer shared at the societal level. The tremendous choice that we have in picking and choosing narratives means that when we partner with people, we generally share a language but not the baseline set of beliefs or values that we would have in an ancestral environment. Historically, shared narratives, or at least the cross-pollination of narratives, kept manipulation in check. Now those systems are breaking down. In the past, those creating and those (others) consuming the narratives—be it religion or myth, news or gossip—had shared fate, and they knew it. Now we live in such a fractured society that most of us have little sense of our shared fate—that we all live, for instance, on a single planet on which we depend. So while it seems that we live in an ever more pluralistic world, in which, for instance, people across religions can intermingle without hatred, our political tribalism has reached a fever pitch, "helped" by algorithms that divide us into silos.

Children are growing up in a world that is designed to hurt them. School, which should be helping young people learn how to be successful adults, is mostly rudderless at best, and actively harmful to development too much of the time. Products and algorithms coming at children will do them harm, their motivational structures will be hacked, their peers will lead them astray. They will not be unscathed. How, then, to become functional adults?

Laboratory of the Self

The "self" is inherently an anecdote, a sample of one. Therefore, the concept of "self as laboratory" will set trained scientists on edge. The problem for humans who are trying to figure out how to live in the world is that we are each our own unique, complex system. There are some universals, to be sure—toxins and advertising and sedentary lifestyles are risky for all of us, many of which we have already discussed in this book. Consider, though, that our internal wiring is so distinct from that of the next person, that for many topics, the advice that works for person A may well fall flat for person B.

To borrow very loosely from Tolstoy, every functional liver is (essentially) the same, whereas every modern human mind is dysfunctional in its own individual way. Your best friend's anxiety, disordered sleep, and perfectionism are not the same as your second cousin's anxiety, disordered sleep, and perfectionism, in either their etiology or their manifestation.

The puzzle of modernity compounds this problem a thousandfold. Humans are capable of inhabiting every human niche that has ever been exploited: we are hyper-plastic. Combine that with a modern environment that is ultra-noisy, and we all face an independent landscape of dysfunction. What this means is that we all have to sort out what works for us as individuals. The advice of other people will vary vastly in its applicability, even when the advice is effective for the person giving it. We have to get good at scientifically testing which things actually result in net positive change within our own, individual, complex systems.

Advice is everywhere. A zillion people have claimed to have figured out how to help us become our best selves. (We are not blind to the fact that, in a sense, with this book we are claiming to have done this as well.) Roughly speaking, such would-be self-help gurus fall into four categories: the con artists, the confused, the correct but of finite applicability, and the universally useful. We posit, and we hope by now you agree, that many evolutionary truths are universally useful.

Effective con artists are hard to see in advance, but it's up to all of us to learn to do so. In the second category are those who are confused, spouting "wisdom" because it attracts people and money, not recognizing that such wisdom may have no relationship to truth or value. Both con artists and the confused are generally playing an entirely social game. Many of the confused, and the con artists, appear to have dispensed with core beliefs altogether, navigating in an entirely social mode with no reference to external reality. Rather than having generated ideas based on their fit with reality, they have generated them based on how they are received by their audience. Sometimes there will be "tells" in how they present material. However you figure out who they are, do not seek such people for advice.

In the third category are people who are correct in claiming that they have discovered something that has worked for them, but (and they may not be aware of this) what works for them may not work for you. Their wisdom has limited applicability. Finally, the fourth category includes those rare people who have advice that is universally applicable.

The trick, then, is in figuring out how to

- Dispense with the con artists and the confused (those in the first two categories).
- Learn how to distinguish, within the third category, between those with advice that works for them but is inapplicable for you; and those who know something that, if you can figure out how to apply it, would improve your life almost instantly. Do this by engaging in a kind of scientific Buddhism. Banish noise, notice small potential patterns, and test hypotheses within yourself for what works.
- Adopt the good advice of those in the fourth category, those few who actually have universally applicable advice.

When the WEIRD world became obsessed with gluten many years back, it looked like yet another fashion trend that would have applicability for a tiny fraction of people. Meanwhile, Bret had been dealing with asthma for decades, using steroid inhalers and other pharmaceuticals on a daily basis, with no end in sight. At the end of his rope—medical doctors had been of no help, except to advise that he try more prescriptions, and that we get rid of all the dust and cats in our lives—he cut gluten from his diet. He didn't reduce it to a small fraction of its former role; he abolished it. Now many years later, not only are his respiratory problems gone, and have been for years, but nearly all of the other small, irritating health issues that he had are gone as well (and while we did try to reduce the dust in our home, we made no such attempt with cats). Does that mean that you, too, would benefit from taking gluten out of your diet? Maybe. Maybe not. It depends on your particular developmental, immunological, culinary, and possibly genetic history, and you will know best by experimenting on yourself. Gluten sensitivity is neither a fiction nor a universal.

The self is subject to the same scientific principles as everything else, with the same kinds of constraints you find when you're trying to study biological phenomena in the field. Complexity and noise are the enemies of signal. The solution involves controlling your experiments as much as is possible given the constraints of the environment. Change only one thing at a time. Do it fully and completely (if you cheat, you've learned nothing, but you may be fooled into thinking that you now have information). And give it time to work.

Types of Reality

Remember Wile E. Coyote, whose life's work was chasing the Road Runner in *Looney Tunes* cartoons? In hot pursuit he often found himself skidding off the edge of a cliff, where he would hang, suspended in air, until he looked down. Gravity did not apply until he recognized that it should. It was funny, because it was ludicrous. It was utterly ludicrous, and yet too many modern people seem to imagine that by changing people's opinions or perspectives, you change underlying reality. In short, they believe that reality itself is a social construct.

We argued earlier that con artists and the confused often operate on a wholly social plane, rather than on an analytical one. How do you avoid becoming someone who assesses the world based on social responses rather than based on analysis, one of those people who are easily fooled by con artists and the confused? Two good strategies are to regularly engage with the physical world and to understand the value of close calls.

The sad truth is that the more "educated" you are today, the harder this is to do. Our current higher education system is steeped in a philosophy that doubts our ability to even perceive the physical world. That philosophy is called postmodernism.[4]

Postmodernists have been at the leading edge of promoting the view that reality is socially constructed. Postmodernism, and its ideological child, post-structuralism, were once contained in a small corner of the academy. These ideologies do contain kernels of truth. They teach us that our sensory apparatus biases us, and that we are mostly unaware of those biases. They reveal that schools, factories, and prisons are similar in their use of power to control populations (as analyzed by Michel Foucault in his metaphorical extension of Bentham's Panopticon). And Critical Race Theory has at its foundation the real observation that the American legal system has had a particularly difficult time emerging from its racist past, and that full recovery from that past is not yet on the horizon. These are a few real and valuable contributions that such ideologies have contributed to the world. But most modern instantiations of postmodernism have jumped the shark.

Sometimes when fringe academic ideas go rogue, they persist longer than they might, but their impact is limited to a few university departments. Not so with postmodernism and its downstream effects. What happens on campus has most definitely not stayed on campus. Postmodernism and its adherents have infiltrated systems far beyond higher ed—from the tech sector to K–12 schools to the media—and are doing considerable harm.[5]

One of the most astounding conclusions of some postmodernists is that all of reality is socially constructed. They have even taken issue with the conclusions of Newton and Einstein, on the basis that the privilege of those scientists is obvious in their equations and, as old white guys, their biases inherently prevented them from knowing anything real of the world.[6] People of particular phenotypes, this ironically biologically deterministic and regressive worldview argues, can't possibly have access to truth.

How do you come to be this confused, to believe that all reality is socially constructed? Have little experience in the real world. No carpenter or electrician could believe that all of reality is socially constructed. No forklift operator or sailor could. Nor, we would have thought, could any athlete.[7] There are physical ramifications of physical actions, and everyone operating in the physical world knows this.

If you have not thrown or caught many balls, or used hand tools, or laid tile, or driven stick shift—in short, if you have little or no experience with the effects of your actions in the physical world, and therefore have not had occasion to see the reactions they produce, then you will be more prone to believing in a wholly subjective universe, in which every opinion is equally valid.

Every opinion is not equally valid, and some outcomes don't change just because you want them to. Social outcomes may change if you argue or throw a fit. Physical outcomes will not.

Everyone, no matter how trapped they are in their body, with its particular flaws and strengths, has the opportunity to experience the world of actions and reactions in the physical world. Not everyone can bike single track, but for those of us who can and do, we face objective reality in the form of roots, hills, and gravity. How, given your particular body, might you force your mind and body into confrontation with physical reality?

Consider this: Our eyes do not produce a static image, like a photograph. Rather, our eyes are tools of our brains, taking note of the world. We are fully embodied—our bodies are not afterthoughts to our brains, or unnecessary to their interpretation of the world. Those eyes, in those skulls, on those necks, atop the torsos and legs and feet that move—it is all part of perception. Perception is an action.[8]

The more you move, therefore, within whatever your particular limits are, the more integrated, whole, and accurate your perception of the world is likely to be.

Movement increases wisdom. So, too, does exposure to diverse views, experiences, and places. We need both freedom of expression, and freedom to explore, because both speak to the value of environments in which outcomes are uncertain. Nature is still available to us. Let us spend time in it, and in so doing generate strength and calibrate our understanding of our own significance.[9]

Humans are evolved to be antifragile: We grow stronger with exposure to manageable risks, with the pushing of boundaries, fostering openness to serendipity and to that which we do not yet know. This is true for both bones and brains. Doing things with nonnegotiable outcomes in the physical world—skateboarding, growing vegetables, ascending a peak—provides a corrective to many wrongheaded ideas currently passing for sophisticated. Some of these include: all of reality is a social construct, emotional pain is equivalent to physical pain, and life is or can be made perfectly safe.[10]

A graduate school mentor of ours, George Estabrook, who was primarily a mathematical ecologist but who also spent many years of field seasons working and living with the practitioners of a traditional agricultural system in the hills of Portugal, wrote this in the introduction to one of his papers:

> It is remarkable how the persistent empiricism of human beings, struggling to make their living in nature, results in practices that make ecological sense, even though they may be codified in ritual or explained in ways that seem superficial or not compelling ecologically. Indeed, local practitioners may have concepts, equally justifiable but very different from those of academics, of what constitutes a useful explanation.[11]

If we were forced to choose between the "useful explanation" given to us by a villager versus one provided by your average academic, about an object that the villager depended on for her sustenance, we would surely choose the villager's explanation. That Costa Rican villager who likely saved our lives by keeping us out of a rapidly rising river, whom we talked about in the Introduction, knew far more about where we were, and how to interpret the signs, than we budding academics did.

You can fool a person, and they can fool you, but you can't fool a tree or a tractor, a circuit or a surfboard. So seek out physical reality, not just social experience. Pursue feedback from the vast universe that exists beyond other human beings. Watch your reactions when the feedback comes in. The more time you spend pitting your intellect against realities that cannot be coerced with manipulation or sweet talk, the less likely you are to blame others for your own errors.

On the Benefits of Close Calls

"When I succeed, it's due to my hard work and intelligence; when I fail, the system is rigged against me, and I had bad luck." It's easy to see the flaw here when stated so clearly, but most adults today are motivated by some version of it in their day-to-day lives. The fact that we tend to believe in bad luck, but not in good luck, makes it more difficult to learn from our mistakes.

When our sons experience setbacks—anything from dropping a glass to slipping on the stairs to breaking an arm—we ask them, "What did you learn?" We also, to their enduring irritation, will often ask them this when they *almost* drop a glass, don't *quite* slip on the stairs, or narrowly *avoid* breaking a bone. They expect it from us now, but in general, both children and adults are incredulous when you ask this question after something has gone wrong. It can be taken as accusatory, rather than sympathetic, and sympathy is what we think we want in the aftermath of an accident or injury. As much as your currently ruffled feathers would enjoy being smoothed, wouldn't you be a more productive and engaged human being if you could learn from what just happened, and thus decrease the chances that you'll experience such a thing again? It is, as we say to our children, about the future. Trying to explain away the past, rather than learning from it and moving on, is a poor use of time and intellectual resources.

Having close calls is part of the set of experiences that are necessary in order to grow up. If your child has been made totally safe, living a life with no risk, then you have done a terrible job of parenting. That child has no ability to extrapolate from the universe. If you, as an adult, are totally safe, you are probably not reaching your potential.

What does "safe" mean, though? When we think about safety, it is tempting to develop a universal rule and stick with it. But this, like everything, is context dependent. Static rules are easy to remember; they are also of little use. Are roller coasters dangerous? Consider the risks of an adrenaline-pumping ride at an established theme park like Disneyland versus one at a carnival. Theme parks have permanence, and the rides have been in place for a long time; they are therefore almost certain to be safer than the frequently deconstructed and rebuilt rides of a traveling carnival.

Similarly, consider the risks of power tools. All blades that are powered by electricity are dangerous, to be sure, and require special attention and practice to be safe around. But if you think that "be careful, it's a power tool" is a sufficient level of warning, you probably aren't knowledgeable enough to be safe yourself. Consider band saws, circular saws, table saws, and radial arm saws: the hazard increases substantially as you move left to right through that list. You are far more likely to have a close call, rather than lose a finger (or worse), if you understand the different risks as you use the different tools.

Finally, consider the risks of a walk in a suburban forest in the United States, versus one in Yosemite, versus one in the Amazon. The risks from the environment are different in each case—other people are a far larger threat in a suburban park than in Yosemite, for instance, while physical injuries may be somewhat more likely in the Amazon and Yosemite than in a suburban park. The primary difference in risk to human health, though, is in the remoteness from medical care in a national park and, even more so, in the middle of the Amazon. As we used to tell our students in advance of study abroad, "Be courageous, but aware of your own limitations, and responsible for your own risks. Assessing risk is a different calculation when you're far from medical help. Lawyers have not gone through the environments we'll be traveling in and made them safe—in that truth lies both much of the fun, and much of the danger, of the journey."

During our eleven-week study abroad trip through Ecuador in 2016, we had one primary and explicit rule: *Nobody comes home in a box.* We had three close calls on and just after that trip. You have already read about the tree fall. In the Galápagos a few weeks later, a remarkable boat accident nearly killed Heather and the captain, and could easily have killed all twelve of us on board, including eight students. It left Heather broken in many ways, nearly incapacitated, but she did not come home in a box.[12] Our students Odette and Rachel were two of those in the boat accident: Odette sustained some injuries, but Rachel was, remarkably, unharmed. Together, they experienced the final close call, which was just half a month later. It was even more dramatic. It warrants a full retelling by them, but here is a précis.

Our thirty students had scattered to research sites to do five-week-long, independent research projects. Odette and Rachel had begun working at a field station in coastal Ecuador, but had gone to the nearest town to make the required weekly email contact with us and to celebrate Rachel's birthday. They had splurged on a second-floor room in the Hotel Royal, which was the tallest building in Pedernales, six stories of unreinforced masonry. Just after they came in from watching the sunset, the room began to shake. They reached for each other and fell together on their knees between the two sturdy twin beds. Then the entire hotel collapsed—both under and on top of them. They were briefly in free fall, along with several stories of cinder blocks surrounding them.

That earthquake on April 16, 2016, was a 7.8 on the Richter scale. It devastated much of coastal Ecuador. Pedernales was at the epicenter and was largely destroyed.

We learned of the earthquake within an hour of it happening. We knew where all of our students were, and only a handful were in the danger zone. We quickly accounted for everyone—everyone except Odette and Rachel. We knew they had gone to a coastal city, which we believed to be Pedernales, for the weekend. The reports out of coastal Ecuador were grim. We talked to Odette's mother several times, trying to reassure her, and to the people who ran the field station where the girls had been—we were assured that the staff who were accounted for were on the move, actively looking. Some of the staff were also missing. Bret began to make plans to go back to Ecuador to look for them. Heather could still barely move from her own injuries from the boat accident, but Bret could go. It wasn't clear that it was the right move, but it was the only action possible, and we needed the girls to be all right.

Mid-afternoon of the following day, after twenty hours of hoping desperately for signs of life, we got a short, grateful email from Rachel. They were alive. They had salvaged almost nothing. But they were alive.

As far as we know, Rachel and Odette were the only survivors from the Hotel Royal. They fell in just the right place—between beds that had been so overbuilt that they withstood several stories of masonry on top of them—so they got lucky. Then their considerable wisdom and clearheadedness pulled them through what would be a nearly twenty-four-hour horror show.

They were trapped under and amid concrete rubble and dust, ghostly apparitions in the light of Odette's tablet, which had somehow both survived and was findable by them in the immediate aftermath of the quake.

Aftershocks began shortly. The concrete slab over their heads moved slightly. They heard voices outside, and they shouted. Three men heard them, and together they all dug through the masonry with their hands, enlarging a tiny gap into one large enough for the girls to get through. Odette had considerable but not life-threatening injuries. As a ballet dancer, she was familiar with pain, but this was far different; she couldn't walk. Rachel was, once again, unharmed.

They needed to get to Quito, but their journey was never simple. They were helped by many good people, and ignored or rebuffed by many who could barely take care of their own. In Pedernales, all was chaos. They saw a woman holding her lifeless child in her arms. They began to hear people murmuring about tsunami. One of their many rides out of town, which seemed promising, fell apart when the driver learned the fate of his family. While ensconced in another ride in the back of a utility van, a makeshift medic cleaned and stitched the long laceration in Odette's foot, an injury that would later require more than one surgery to heal. One of their rides ran out of fuel. Another had to turn around at a missing bridge. Again and again they were returned to Pedernales, the scene of twisted concrete and sobbing people and a fine white dust on everything—in part, the remnants of the Hotel Royal. At last, having found seats on a bus to Quito, they encountered massive landslides from the earthquake that nearly blocked the road. As they edged past, chunks of earth disappeared into the chasm below.

Finally, they made it to Quito. They were alive, and they were safe.

Nobody came home in a box on that trip. And despite the extensive physical damage and psychological trauma she endured, Odette later said to us, "The trip was singular, remarkable, terrifying, extraordinary. Even if I knew everything that was going to happen to me, I would still go. It was so important to me."

On Fairness and Theory of Mind

Many people are failing at adulthood in exactly the domains in which they are supposed to be adults. When Evergreen, the college at which we were tenured and which we loved, went haywire in a flurry of strategically named "social justice" actions, almost no adults stood up. To most of the world that paid attention, it appeared that a ragtag bunch of entitled college students were forcibly taking over a college, and that is, indeed, part of the story. A more accurate yet still woefully incomplete rendering is that, behind the scenes, a few faculty bullies had indoctrinated students and taken over several key college functions; and the administration—the people who are paid to act like adults when things at a college go haywire— evidenced dereliction of duty.[13] Being an adult, in part, means not abdicating responsibility, especially when others are depending on you.

Being an adult also means engaging in cooperation on a number of levels. We may engage in kin selection (in which we preferentially help our relatives), direct reciprocity (*I help you raise your barn, or move into a new apartment, and you will help me later*), and indirect reciprocity (*I do a good deed publicly, which enhances my reputation*).[14] It is rare that we are conscious of these theoretical considerations as we act, of course. Our morality extends from a fluid mixture of these forms of cooperation. Much of the variation within groups over time, in terms of commitment to other group members and to the success of the group, can be explained in terms of group stability.[15] When your group is threatened, you rally, and the bonds within the group grow stronger. In good times, however, when things are easy, group stability tends to fray, first around the edges, and ultimately at the core. Again, economic markets prey on this tendency, destabilizing our sense of self and community, causing us to look elsewhere for the missing ingredient that will, finally, make us happy, productive, and secure.

We humans are particularly adept at recognizing that our perceptions of the world are not shared by everyone. This ability to recognize that other people understand the world differently is theory of mind, which we have already invoked several times.

Organisms with theory of mind have the ability to distinguish between subject and object. Baboons in the Okavango Delta in Botswana, for instance, know the difference between "She threatens my sister" and "My sister threatens her." They exhibit the first glimmers of theory of mind—the ability to track not just what their own model of reality is, but also that of other individuals, even when those models differ from their own.[16] We can also deduce that all the usual suspects—the wolves and the elephants, the crows and the parrots—have theory of mind, given their social, long-lived, multigenerational-families and high parental caregiving ways.

One thing that theory of mind provides potential access to is a sense of fairness. The concept of what's "fair" didn't originate with philosophers. It didn't emerge with city-states, or with agriculture. It wasn't new to hunter-gatherers, either, or to our first bipedal ancestors. Monkeys keep track of what's fair, and what's not, and they have decided opinions about unfair practices in their social realm.

Capuchins—New World Monkeys that live in large social groups—will, in captivity, barter with people all day long, especially if food is involved. *I give you this rock and you give me a treat to eat.* If you put two monkeys in cages next to each other, and offer them both slices of cucumber for the rocks they already have, they will happily eat the cucumbers. If, however, you give one monkey grapes instead—grapes being universally preferred to cucumbers—the monkey that is still receiving cucumbers will begin to hurl them back at the experimenter. Even though she is still getting "paid" the same amount for her effort of sourcing rocks, and so her particular situation has not changed, the comparison to another renders the situation unfair. Furthermore, she is now willing to forfeit all gains—the cucumbers themselves—to communicate her displeasure to the experimenter.[17]

Markets prey on our sense of fairness. They fool us into thinking that everyone else is getting grapes, while we are stuck with cucumber. If other people already have those better things, why don't we? Our sense of fairness is thus kept off balance, always threatened by the invisible other consumers who already have the next big thing, and thus must be doing better than we are. We are still trying to keep up with the Joneses, but the Joneses are no longer our neighbors. They are now a tiny fraction of the world's elite piped into our screens, and photoshopped to boot.

As humans, one of the ways that we test the moral waters and assess the mood of a group and its boundaries is through humor. It helps mitigate questions of fairness. Humor is the mechanism by which we sort out the gray area of what can and can't be said. A humorless society, community, or group of friends likely has large problems lurking just beneath the surface. Furthermore, attempts to induce laughter inorganically—as with laugh track—is the market once again trying to intrude on honorable human tendencies to bond over shared experience and understanding. Laugh track ultimately renders us *more* humorless, and less capable of connecting with actual human beings.

On Addiction

Many things have a pathological version. Pathology is not the same as "downside"—senescence is a downside of early adaptive traits, but it is not pathological. In contrast, arrogance is pathological confidence.

Positive obsession has many words in English: passion, focus, drive. The primary manifestation of negative obsession, of pathological obsession, is addiction.

Obsession is agnostic with regard to whether the thing obsessed over is healthy or unhealthy. You can obsess over a love interest, which might lead to the love of your life. You can obsess over a particular varietal of mango, which may cause you to spend more time than strictly necessary seeking out such mangoes. You can obsess over what color to paint your walls, what order the paragraphs should go in, whether to tell your friend that her husband is a lout.

Addiction is one end point of an unhealthy obsession.

One common misapprehension of addiction is that if you use an addictive substance, you will become addicted. Consider heroin. It seems to be true that adding exogenous opioids to your body renders you less capable of producing them endogenously—from within, by the body itself. Therefore, when the exogenous molecule disappears, because you ran out of money or your dealer got arrested or you went into rehab, it is painful because you no longer have the ability to produce *endo*genous opioids. One

might conclude that everyone who uses heroin and other exogenous opioids therefore tends to become an addict.

And yet we know that most people who try drugs don't become addicts.[18]

Give rats a lever that provides amphetamine on demand, and sure, they
press the lever. If there is nothing else available to them, they become addicts. Give them an enriched environment, though, with lots of other cool
rat stuff to do, and they don't become addicts. They do other good rat stuff
rather than become addicts.[19] Perhaps they are, in fact, freed up to become
obsessed over something that is healthy.

What is "healthy," of course, is ever more challenging to decipher,
in no way helped by the presence of market forces in almost every deci-
sion. Attempting to understand humans as an evolutionary phenomenon, as we are doing in this book, assumes that all of our minds are,
behind the scenes, doing a cost-benefit analysis between choices that we
perceive that we have. From how to walk, to whom to mate with, to what
book to read, it's all cost-benefit analysis with the target of increasing
fitness. Our software is built to maximize our fitness, even if our conscious minds have other priorities. But our software has an increasingly
difficult time telling signal from noise, because our map of what enhances fitness in the ancestral world does not prepare us well for the
modern world.

Our intuitive sense of the fitness value of behaviors is thus often wrong
in modernity. Our intuition had a greater chance of leading us to the right
choice before the Industrial Revolution, before hyper-novelty was ubiquitous. Many of us are now effectively able to pull levers, like rats with access
to amphetamine, and get a concentrated burst of euphoria that doesn't just
obscure the risk of that euphoria, but makes it ever less likely that we can
turn away from it in the future. It's another instantiation of the Sucker's
Folly: the reward obscures the cost.

Every drug, or other potential object of addiction, creates a level of
reward that varies with other parameters. "Reward" is not binary—it is not
simply a positive or a negative. The valence and size of the reward depend,
in part, on what the other possibilities are—the opportunity cost. Should I

pursue that person as a mate? Should I smoke a joint? Should I binge-watch the newest Netflix series? Should I browse social media? Those aren't complete questions until you know what you would be forgoing in order to be with him, smoke the joint, watch the show, or engage online. That is, the cost-benefit analysis is incomplete until you compare it to what else you could be spending the time on.

What the rat enrichment experiments suggest is that one contributor to addiction can be boredom. Or more specifically, a lack of awareness, or obfuscation, of opportunity cost. Boredom is effectively synonymous with the "opportunity cost" having gone to zero: if you believe there is nothing else enriching that you could spend your time on, then the calculation of whether or not to engage with a particular substance or action is skewed, particularly if that substance or action results in a feeling of enrichment, even a false one.

It is, of course, too simple to say that boredom causes addiction. There are many factors at play: the limiting nature of ancestral environments did not require self-regulation for most substances or behaviors; both trauma and psychological disorders disrupt decision-making processes; emotions are hijacked by addictive substances and behaviors that create a false incentive structure; and social pressures often drive calculations toward consumption.

All of this is part of the equation. What is interesting and possibly instructive is that all of the factors just listed effectively skew the cost-benefit analysis and obscure our understanding of opportunity cost. Boredom, as a proxy for opportunity being zero, seems to be a through line in the story of addiction.

We take advantage of our vulnerability by creating systems that evolve to be addictive. Social media is an excellent example.[20] In retrospect, we should not be surprised that we created a system that addicted even its creators. In the future, we should be far more careful about opening up Pandora's boxes. And we should create—and encourage the creation at larger societal scales—new opportunities for engagement, for creation, for discovery, for activity that provide an alternative to the boredom that leads to addiction.

The Corrective Lens: How to Give Yourself a Raise

- **Explicitly aim to be an adult.** Do this, in part, by regularly asking yourself the questions that we posed at the beginning of the chapter (Am I taking responsibility for my own actions? Am I being closed-minded? . . .), and by minimizing the effects of economic markets on your daily life.

- **Become aware of the constant flow of information telling you what to think, how to feel, how to act.** Do not let it into your mind. Do not let it steer you. Your internal reward structure needs to be independent and ungameable. That independence, in turn, should allow you to collaborate well with others who are similarly independent. Be wary of those who may well be nice, but who are captured.

- **Always be learning.** Look for collaborators. Play at competition, and be prepared to stop playing if things get real. Be skeptical, if not suspicious, of any novel prescription for which the rationale is unstated or thin on reflection.

- **Revive, or create, rites of passage in your life.** Celebrate not just the passage of time (birthdays, holidays), but also developmental transitions. Honor graduations and marriages, births and deaths, but also career and job changes and promotions, the completion of important analytical or creative tasks, and the ends of eras, when they are recognizable as they end.

- **Seek out physical reality, not just social experience.** Pursue feedback from the physical universe, not just from subjective social sources. Move your body. Gain experience with model systems that tell you how things actually work.

- **Get over your bigotry.** Variation is our strength. Not just sex and race and sexual orientation, but class, neurodiversity, characteristics of personality—all of this adds to what we can accomplish on Earth.

- **Place equality where it belongs.** Equality should be focused on the equal valuation of our differences. It should not be a bludgeon for uniformity.

- **Smile at people**—the people with whom you live, the person behind the counter, the stranger on the street.
- **Be grateful.**
- **Laugh daily, with other people.**
- **Put your phone down.** No really, put it down.
- **Define your fights for whom and what you love**, rather than against whom and what you hate. If a mob ever comes for people you know, people whom you consider friends, stand up and say, "No, you're wrong." Be honorable and courageous when bullies move in. Speak up for what you know to be true, even if it makes you a social pariah.
- **Learn how to give useful critique without backing the other person into a corner.** With our children, when they fall off a unicycle or don't do well on a math test, we tell them it's "not your best work." It's true; it doesn't pretend that every action is worthy of a gold star; and it demonstrates that we know that they can do better work, and that this wasn't it.
- *Count* **fewer things about your life (calories, steps, minutes), and** *do* **more.**
- **Develop a theory of close calls.** When a close call occurs, have a plan for how you will leverage it to gain a better understanding of yourself and the world. Calm down, and level up.
- **Learn to jump curves.** Diminishing returns are a factor for every complex phenomenon, so learn to jump curves (put another way: consider learning a *new* thing rather than being a perfectionist and trying to get ever better at whatever you are already really, really good at). We will speak to this more in the final chapter.

Chapter 12

Culture and Consciousness

ON THE SHORE OF A LAKE ON ORCAS ISLAND, IN THE FAR NORTHWEST corner of the United States, there is a campfire. It's a crisp October night, and the sky is both clear and dark enough to see the stars. Many in our class are sitting around the fire. A couple of students have guitars, one a harmonica, so there is music. Sometimes it dominates, sometimes it weaves around the conversation. We are warming our bodies and sharing ideas, memories of the day. We speak of research designs that different groups came up with to answer the question, Does biodiversity vary with altitude on this island that we are on? It is a question that must have concerned people here millennia ago, even if not in those terms—hunters and gatherers would have kept track, subconsciously if not consciously, of where they were most likely to be successful in their attempts to find food. We speak of sex and of drugs. How should we view sex without commitment? If the use of hallucinogens is adaptive, should everybody do them? We speak of staying warm.

We have sat around many campfires over the years. Hopefully you have, too.

The Age of Information brings the promise of a collective (metaphorical) campfire, a decentralized thing where people who have never met in real life can be warmed by the presence of other minds, sharing ideas and reflections.

But the online world, though it holds promise, does not have the structures that made discussion around the hearth so valuable. An ancestral campfire places everyone's reputation—earned over a lifetime—front and center. Around an ancestral campfire, each person would have some basis for elevating or discounting claims and proposals based on the individual's known strengths and deficits, and taking the history of the discussion into account. The virtual campfire is, by contrast, a free-for-all. We don't really know one another, our visible history is often misleading, many users are anonymous, and some participants have a hidden dog in the fight. The list of flaws is immense. Traditional campfires are waning in frequency, and virtual campfires often bring new problems; are there other ways to bring about a campfire renaissance? We need one. Campfires both metaphorical and literal are a convergence point for culture and consciousness, where people come together in good faith to learn the old wisdom and to challenge it.

Let us begin with definitions. These will not precisely match others' definitions, but it is important, on this topic, to have stated what we are talking about. For our purposes:

Culture we define as beliefs and practices that are shared and passed between members of a population. These beliefs are often *literally false, metaphorically true,* implying that they result in increased fitness if one acts as if they are true despite the fact that they are either inaccurate or unfalsifiable. Culture is a special mode of transmission because it can be passed horizontally, rendering cultural evolution immensely faster and more nimble than genetic evolution. This also renders culture noisy in the short term, before new ideas have endured the test of time. Long-standing features of culture, by contrast, constitute an efficient packaging of proven patterns. Culture can spread horizontally, but its consequential parts are ultimately passed vertically, from generation to generation. Culture is received wisdom, generally handed to you by ancestors, and efficiently transmitted.

Consciousness we define—as we laid out in the very first chapter of the book—as that portion of cognition that is newly packaged for exchange,[1] meaning that conscious thoughts are ones that could be delivered if someone asked what you were thinking about. It is emergent cognition, where innovation and rapid refinement occur. Conscious thoughts may never be conveyed, but they can be, and the most important ones are, as consciousness is most fundamentally a collective process in which many individuals pool insights and skills to discover what was previously not understood. The products of consciousness are, if they prove useful, ultimately packaged into (highly transmissible) culture.

We have said before in the book that the human niche is niche switching. More specifically, we argue that the human niche is to move between the paired, inverse modes of culture and consciousness.

As an example, let us consider the Nez Perce people, who have lived in the Pacific Northwest for many thousands of years. Since they arrived, they have inhabited a rich land, and they now have well-established cultural rules that keep them safe and thriving. Their diet has long included bulbs—the storage organs of plants, which do not want to be eaten. On these lands where the Nez Perce came to live, both camas (with highly nutritious bulbs) and death camas (with bulbs that are toxic) grow. When not in flower, these bulbs are incredibly difficult to tell apart. The Nez Perce may not have been the first on this land, but someone was, and those first people could not benefit from names that would make the danger clear. Yet they learned the distinction—presumably through trial and error. It was likely a messy, tragic process. By the 19th century, however, when Spaniards were documenting what they saw among the Nez Perce, the system of distinguishing nutritious camas from their deadly relatives was nearly perfect. This is culture.

When humans are exploiting a well-understood opportunity, like 19th-century Nez Perce distinguishing camas from death camas, culture is king. But when novelty renders ancestral wisdom inadequate—as it was for the more ancient, ancestral Nez Perce upon arrival in the Pacific Northwest—we need to shift to consciousness. Through parallel processing of multiple human minds, our consciousness can become collective, and we can solve problems that neither we could solve as individuals nor our ancestors could have even imagined.

Put another way:

In times of stability, when inherited wisdom allows individuals to prosper and spread across relatively homogeneous landscapes: *Culture reigns.*

But in times of expansion into new frontiers, when innovation and interpretation, and communication of new ideas, are critical: *Consciousness reigns.*

That said, novel levels of novelty, such as we are experiencing now, are a special danger. This means that what's needed today—and urgently—is a call to consciousness on a scale that we have not seen before.

Consciousness in Other Animals

In other animals, when social, species-level generalists have wide geographic range, individuals will often become problem-solving specialists that share their insights with conspecifics. This is true in humans, and in all the usual suspects: wolves and dolphins, crows and baboons, and so on. These animals can be understood to have a form of consciousness.

Tree frogs, octopuses, and salmon do not, however, have consciousness. These three clades vary widely in life history and intelligence—octopuses are famously smart, and excellent at solving puzzles; tree frogs and salmon, while fascinating, are not capable of the kinds of cognitive feats that octopuses are. What all of these clades have in common is that individuals are not social.

A large congregation of western chorus frogs on a Michigan pond on an early spring night, loud though it may be, is not a social group. The frogs come together to mate, but once the deed is done, they separate from one another and never interact again. Chorus frog parents never even meet their offspring. Similarly, salmon swim upstream en masse, and spend time near one another competing for the best nest sites—but aggregation is not the same as sociality.

It's the difference between the people on your subway car (with whom you aggregate), and the people with whom you share a house (with whom you are, under most circumstances, social). Even this is a flawed example though as, being humans, we do notice and remember certain people on the train, especially if we see them every day, or if they strike us as intriguing, even if we never exchange words. Aggregating is just coming together in space. Subways are functionally human aggregators, but they also bring us into social contact, in part because humans are always on the lookout for social opportunities. A train full of tree frogs would not become social, no matter how often they commuted together.

By comparison, a large congregation of baboons on the Okavango Delta has lasting power—there are multiple hierarchies that predict who will get to eat first and whose baby will thrive, and the baboons keep track not just of individuals, but of the relationships between them.[2] Their culture is evolving, just as ours is.

Sociality involves recognition of individuals, the tracking of social fate, and iterated interactions that are, at least plausibly, continuing into the future.

Innovation at the Margins of the Ancestors' Wisdom

During the peopling of the New World, when was relying on consciousness more effective than relying on culture? Under what circumstances are cultural rules more trustworthy?

As the Nez Perce or their ancestors moved into the range of camas and death camas, they were looking for food in an increasingly unfamiliar landscape. The staples they had come to know were their cultural standbys. As those familiar foods became harder to source, innovation became ever more necessary. They were reaching the limits of their ancestors' wisdom, and confronting a puzzle for which the best tool was consciousness.

As a people move across space, it is relatively easy to notice as the ancestors' wisdom becomes less applicable. As a people move through time, however, as we all do, elders may not recognize their wisdom becoming out of date. The young see it. It is no accident that those who are coming of age in times of change push boundaries, and that language and norms change somewhat with each generation. Throughout history, the ancestors' wisdom has generally remained relevant long enough for new generations to get their footing, to know what needs to be pushed against. As a people move through time that is changing extremely rapidly, however, as our world is now, it is more difficult to know what to do with the increasing irrelevancy of the ancestors' wisdom, and with what to replace it. The margins of the ancestors' wisdom are rarely hard and fast. At those margins, wherever they are, it is time to niche switch.

Consider three broad contexts in which humans have learned and innovated in times past. The first is the utterly new idea: the idea that springs to mind often unbidden and without explanation. This was the territory that the first Mayan, Mesopotamian, and Chinese[3] people were in when they innovated farming. Similarly, the innovation of the wheel, metallurgy, and pottery. Before those things existed, nobody knew they were possible. The second context in which innovation occurs is when you know that something is possible, on the basis that it's been done before, but you have no idea how to make it happen. The Wright brothers saw flight in other organisms, and felt confident that it could be accomplished by machine. Third and finally, you might have instruction: you know what you're shooting for, *and* have someone or some set of rules or instructions telling you how. Between school and YouTube, we often conflate this third kind of learning for the only kind of learning that is possible. The third type of learning is the most cultural; it is the learning of received wisdom. In contrast, humans are at our most conscious, and therefore our most innovative, in the first two contexts.

When the status quo is no longer sufficient, we must seek to innovate, to push beyond how it's always been done. The status quo is in inherent tension with our unique insights. Those ideas that we have deep at night are often syntheses, reflecting the pulling together of common threads into uncommon meaning.

Conformity

In 1951, social psychologist Solomon Asch asked to what extent social forces alter people's opinions. Like baboons, we definitely track what other people think. But to what extent does knowing what other people think change what we ourselves proclaim?

In what is now considered the classic experiment on conformity, Asch asked people a simple, factual question: Which of three lines is the same length as a fourth line? The question wasn't difficult, and the answer wasn't ambiguous. When, however, a "naive" participant was put in a room with several "confederates"—people in on the ruse—and those confederates provided identical wrong answers, only one quarter of the naive participants always resisted social pressure and answered correctly. The vast majority of naive participants succumbed to social pressure sometimes (although only a tiny fraction of participants gave the wrong answer every time).[4]

Unlike many classic psychology experiments, Asch's has stood the test of time. It has been widely replicated, under a variety of conditions. Among other things revealed in the decades since Asch first did his work in the mid-20th century, women are found, in some studies, to conform at higher rates than men[5] (which is in keeping with women being more "agreeable"). Conformity has a time and a place—like most traits, it is not simply worse (or better) than not conforming.

There is a tension between conforming and disagreeing in the face of apparent inconsistency. This tension is a hidden strength of humans—the push and pull between wisdom and innovation, between culture and consciousness.

Humans are generalists at the species level, but have tended to be specialists at the individual level. Historically, we have combined forces in social groups, such that in a single group, many people with distinct skills created an emergent whole, one in which generalist capabilities emerged even if all members of the group were specialists. Now, though, it is time to innovate, because change is accelerating, and the received cultural wisdom isn't sufficient. Individuals themselves becoming more generalist— through learning skills across domains, for instance, rather than diving deep into only one—will help us in this endeavor.

It is important to *know* what the group thinks, but that is not the same as *believing* or reinforcing what the group thinks. In a time of rapid change in particular, then, it is important to be willing to be the lone voice. Be the person who never conforms to patently wrong statements in order to fit in with the crowd. Be Asch-Negative.

Literally False, Metaphorically True

Cultural beliefs are often literally false, but metaphorically true.

Consider farmers in highland Guatemala who have a long-standing tradition to both plant and harvest crops only when the moon is full. This, they say, allows the plants to grow stronger and resist insect damage. What possible protective capacity could the phase of the moon have on crop health? Presumably none. But the phase of the moon *can* synchronize the farmers. A full moon is effectively a giant sky clock, a keeper of time that everyone in the region can see. If all farmers in the region believe that a full moon has salutary effects on their individual crops, they will likely restrict planting and harvesting to the full moon—and this will, in fact, benefit everyone's crops, just not for the reason the farmers believe. A belief in the power of the moon to directly affect crops effectively satiates predators,[6] by concentrating the harvest into brief periods, during which time crop predators cannot eat all of everyone's crops.

It is easy to dismiss many myths and beliefs of old, precisely because they are literally false. Indeed, doing so is almost a sport among some hardheaded people. Take astrology. It is clearly beyond reason to imagine that the stars that we see, many of which are thousands of light-years away, are having a direct impact on human behavior. Similarly, it is beyond reason to believe that a passel of angry gods is the reason for tsunamis, yet among the Moken, those who believe in those gods survive at higher rates than those who don't. And it is surely beyond reason to believe that a full moon is protective of crop health, yet among Guatemalan farmers, precisely that belief results in more productive farming.

In each case, the belief is literally false, but metaphorically true.

This means that the cover story isn't true, but when people behave as if it were, they prosper. This is how religion and other belief structures spread. Even if such things are not literally true, acting as if they are benefits people; sometimes it even benefits the biodiversity and sustainability of the land on which they live.[7]

In its modern, tabloid form, astrology is bunk. But astrology has probably not been so everywhere and for all time. *If*—and this is a big *if*—you control for where a person was born, might not the time of year that they were born have effects on how they develop, and therefore who they become? And aren't astrological signs just an ancient way of keeping track of the months, more or less? If we look at astrology this way, rather than as a modern indulgence that is too free of context and history to have meaning, it begins to look promising. Is a newborn in a Minnesota winter exposed to the same pathogens and activities as a newborn in a Minnesota summer? Surely not.

And sure enough, there has been work done to bear out this idea: culling data from over 1.75 *million* records at New York-Presbyterian/Columbia University Medical Center, for people born between 1900 and 2000, researchers found clear correlations between birth month and lifetime disease risk for more than fifty-five different conditions.[8] With affected systems ranging from cardiovascular to respiratory, from neurological to sensory, the sheer number and breadth of medical conditions that vary in lifetime risk by birth month should be enough to make a thoughtful person rethink a wholesale rejection of careful astrological thinking.

For if there are demonstrable differences in disease risk by birth month, why should we imagine that there are no differences in personality?

As an aside: One prediction of this approach to astrology is that, if you include both birth place and date, astrology will have less power to predict lifetime disease risk the closer you get to the equator, where seasonality is much reduced from that in the temperate zones. Another prediction is that, the more a person moves around as a child, the less predictive astrology will be for them. (And if you don't include birth place, astrology should have no predictive power at all.)

Distortions that help you survive and thrive are adaptive. Myths and taboos often make little sense to outsiders, and some of them are surely misguided, even counterproductive for those who honor them. Some surprisingly precise taboos are likely overgeneralizations from an actual event. Among the Camayura of the Brazilian Amazon, the eating of scaleless fish is forbidden for both pregnant women and their husbands.[9] It may well be that, long ago, a terrible fate befell a woman, her unborn child, or her entire family, after eating a fish without scales, and that the fish was the only explanation that stuck. Similarly, on the haut plateau of Madagascar, in the village of Mahatsinjo, there is a taboo against eating hamerkops, a close relative of pelicans. This taboo is directly tied to villagers having seen one fly over just as a man died.[10] Elsewhere in Madagascar, it is taboo for young men to eat mutton before wooing; taboo for pregnant women to eat the meat of hedgehogs, or to walk through fields of pumpkin; taboo for a son to build his home to the north or east of his father's house.[11] To our Western sensibilities, this seems like superstition, pure and simple.

The word for taboo in Malagasy—*fady*—has a complex meaning as well. In Betsimisaraka, the language of the people of northeastern Madagascar, *fady* means both taboo and sacred.[12] That which is *fady* is mandated by the ancestors—be it mandated that you don't do it, or that you do.

Despite the preceding examples, many beliefs, myths, and taboos are literally false, metaphorically true. Malagasy *fadys* come cloaked in the language of gods and ancestors, but it is still easy to see the wisdom in many of them if you simply look at the prohibition: Do not build a house over or against a new landslide. Do not step on a dead dog, as you might get hydrophobia (rabies). Do not divorce your wife while she is pregnant.[13] We predict that those taboos that have lasted the longest are most likely to be hiding an important cultural truth in plain sight. Beware Chesterton's *fadys*—the old ideas may have hidden truths, and those truths may be difficult to recover once they have been dismissed.

Joseph Campbell observed that "mythology is a function of biology."[14] He was correct. As an evolved creature you are built to succeed, and sometimes that involves telling yourself stories. Finding yourself in a raft near the top of a dangerously tall waterfall, you might be about to die. If you believe that the shore is within reach, and paddle like hell, you just might make it. Those deflated by long odds will leave no trace. Belief can be the difference between life and death.

Religion and Ritual

All cultures have ritual. Death rituals are ubiquitous, birth rituals nearly so. Some rituals are rites of passage to celebrate new babies, coming of age, marriage. There are rituals—traditions, perhaps, given their reliably repeating nature—to celebrate the first planting of the year, and the harvest, and astronomical events like the solstices and the equinoxes. As we have come to live in larger and larger groups, surrounded by ever more anonymity in our daily life, regular holidays, with their attendant shared cultural norms, help to keep us in sync, to act as though we are in fact part of something larger than ourselves. Rituals, which are not inherently religious but have a strong tendency to be so, often include food, music, and dance.[15]

Ritual and religious devotion are demonstrably expensive. Not only do most cultures spend a substantial fraction of their resources and time on structures and ceremonies intended to impress a cold and indifferent universe, but religions expend a great deal of social capital telling believers what they are not allowed to do. If anything dwarfs the cost of religion, it is the opportunity cost of religion. Were it true that religion was maladaptive, these huge costs would constitute a major vulnerability for faithful populations. Atheists who behave just like these believers, except for the fact of skipping religiosity and reinvesting the massive dividend, should displace them as a regular feature of history. If religiosity had no adaptive benefit, great leaders in every population's history would have said, "All you must do is work hard and ignore their mumbo jumbo and their lands will be yours." But that's not what we find. Instead, we find great leaders saying things about God and his quirks, his preferences and his plan for us. Why is that?

Religiosity is adaptive,[16] and moralizing gods, while not being a prerequisite for the evolution of social complexity, seem to help sustain multiethnic empires once they have become established.[17] As moderns, we are often eager to throw off the spiritual and religious chains of the past, but beware Chesterton's gods. Religion is an efficient encapsulation of past wisdom, wrapped in an intuitive, instructive, and difficult to escape package.

Sex, Drugs, and Rock 'n' Roll:
On the Sacred versus the Shamanistic

Culture is in tension with consciousness, much as the sacred is in tension with the shamanistic. The sacred is to culture as the shamanistic is to consciousness.

The sacred is the reification of received religious wisdom, the sine qua non of a particular religious tradition, that which has stood the test of time and proved valuable enough to the ancestors to be passed on as holy. That which is sacred has a low mutation rate—it changes infrequently—and is highly resistant to change; it is built for a static world. The sacred is protected from corruption (or at least, it's supposed to be), and is often insulated from the corrupting influences of secular power, wealth, and reproduction. The orthodoxy of the sacred exists in persistent tension with the heterodoxy of the shamanistic.

The shamanistic is high risk, high creativity. It has a high mutation rate, and therefore a high error rate. It explores a huge number of new ideas, most of which are poor. It challenges orthodoxy—that which is sacred. The shamanistic is practically mandated to explore and play with cultural norms. It does this in a variety of ways, such as through altered states of consciousness, which include dreams, trances, and use of hallucinogens.

Expanding consciousness through hallucinogens is a widespread phenomenon. In the Huichol people of central Mexico, whose ancestors arrived in their lands at least fifteen thousand years ago, small groups make annual pilgrimages across hundreds of miles of rough terrain to find and ceremonially ingest peyote. Every Huichol hopes to make the pilgrimage at least once in a lifetime.[18] Among the Tarahumara of northwestern Mexico, shamans ingest several species of hallucinogens, in search of the evil beings that have brought illness, but so too do long-distance Tarahumara runners, who are both warding off evil and finding strength in the drugs.[19] In nearly every known culture there is use of something, be it strictly hallucinogenic or not, that breaks a person out of the normal, everyday experience and allows for a different perspective to emerge. This is consciousness revolutionizing culture.

When the ancestral wisdom runs out, humans pool their dissimilar experiences and expertise to discover how to bootstrap some new way of being. Identifying when the ancestral wisdom has run out in a particular domain is tough, and there will always be tension between those who want to stay the course and those who are looking to break with tradition and try a new way. Functional systems need those advocating for both—for culture and for consciousness, for orthodoxy and heterodoxy, for the sacred and the shamanistic.

The Corrective Lens

- **Sit around more campfires.**
- **Honor or create rituals that recur**—annually, seasonally, weekly, or even daily. They might be ancient and religious in origin (e.g., honoring the Sabbath or Lent—a time for both selective privation and community), astronomical (e.g., recognizing and celebrating the solstices and equinoxes), or entirely new with you and yours.
- **Be Asch-Negative.**
- **Teach children how to bootstrap their own program**, so that they can be individually conscious. The tension that we have described between culture and consciousness has an analog during development. Trying to teach children precisely how to be adults, by inculcating them solely with the cultural rules that have come before, will fail. In a world of hyper-novelty, many aspects of culture are ever less relevant, and consciousness is imperative.
- **Consider engaging with psychedelics, carefully,** if there is anything in you that is curious. They are now legal in some places. But consider engaging them as the powerful cognitive tools that they are, not as a form of recreation. That doesn't mean you can't have fun.

Chapter 13

The Fourth Frontier

HUMANS MAKE SENSE OF THE PAST AND IMAGINE THE FUTURE. WE HAVE help, in this, from our uncommonly large frontal lobes and from one another. Our children are exceptionally curious, and learn from adults and from one another, from the environment, and from experience. We coalesce in large groups, multiple generations working and living side by side. We use language, experience menopause, mourn our dead, and have rituals to mark events and seasons. We harness the productivity of earth, sea, and sky for our own ends. We domesticate other organisms for food and textiles, labor and transport, protection and friendship. We tell stories, both fact and fiction. We have unlocked many of the universe's secrets, substantially freeing ourselves from the natural order that created us.

But many of our strengths are also cryptic weaknesses. Our outsize brains are prone to confusion and miswiring. Our children are born helpless, and they remain dependent on us for an uncommonly long time. Our great linguistic diversity severely limits to whom we can talk. Even our bipedal gait, so important in allowing us to move and carry things on the ground, comes with risk to mother and baby in childbirth, and reliably causes back pain. We're gossipy, sentimental, and superstitious. We build extravagant monuments to fictional gods. We are arrogant and confused, often mistaking the unlikely for the inevitable, even as we downplay massive and obvious hazards. In everything, trade-offs.

Creatures seek untapped opportunity, and they exploit it. Successful exploitation of novel opportunities temporarily raises the limit on the number of individuals that can live in a given habitat—it produces times of comparative plenty, when births outnumber deaths, and populations rise to their new carrying capacity. "Times of plenty" *is* economic growth. When the usual order is restored as births and deaths are once again balanced, we hit equilibrium, and life becomes harder again. Growth feels good, and it is not surprising that we are obsessed with it. It is adaptive to be obsessed with it. Or at least it has been until now.

Our obsession with growth creates two problems. The first is that we have convinced ourselves that growth is the normal state and that it is reasonable to expect it to go on and on. That patently ridiculous idea—exactly as hopeful and deluded as the search for a perpetual motion machine—causes us to stop searching for other possibilities. While this expectation greatly reduces the chances that we will miss out on growth, it also prevents us from recognizing and pursuing more sustainable options. Second, because we regard growth as normal rather than exceptional, we behave destructively to feed our addiction.

Sometimes we violate our stated values by inventing justifications to steal from a population that has resources but not the means to defend them. Other times we degrade the world, and inflict decline—the opposite of growth—on our descendants in order to fuel current expansion. The former scenario accounts for many of the greatest atrocities in history. The latter explains the modern experience of watching the goodness of our planet liquidated before our eyes. *Growth über alles* is a disastrous creed.

Humans thrive in almost every terrestrial habitat on Earth. We are a broadly generalist species, with highly specialized individuals, who have shape-shifted and niche-shifted into nearly every environment on the planet. This has meant interacting with frontiers, over and over and over again. Here we describe three types of historical frontiers: geographic, technological, and transfer of resource. Then we will propose a fourth.

Geographic frontiers are what we tend to think of when frontiers are invoked: the vast unspoiled vistas, the abundant and yet uncounted resources. All of the New World—North and South America, the Caribbean, and every island near the coasts—was a vast geographic frontier for the Beringians. The frontier of the New World was fractal, so the descendants of the first Americans discovered even more: To the Ahwahneechee Indians, Yosemite Valley was a geographic frontier. To the Taino, the Caribbean was a geographic frontier. To the Selk'nam people of far southern Chile, Tierra del Fuego was a geographic frontier.

Technological frontiers are moments when innovation allows a human population to make more, or do more, or grow more, than they did before the innovation occurred. Every human culture that has terraced hillsides, decreasing runoff and increasing crop production, was confronting technological frontiers—from the Inca in the Andes to the Malagasy on the haut plateau of Madagascar. The first farmers in China, Mesopotamia, and Mesoamerica were doing so, and the first ceramicists—who dug clay, formed it into useful shapes, and fired it in coals—were doing so as well.

Finally, there are transfer of resource frontiers. Unlike geographic and technological frontiers, transfer of resource frontiers are inherently a form of theft. When people from the Old World came across the Atlantic and landed in the New World, they may have at first imagined that they had stumbled upon a vast geographic frontier, but they hadn't. In 1491, the New World is estimated to have had between fifty million and one hundred million people in it, with uncountable distinct cultures and languages. Some people were living in city-states, among astronomers, craftsmen, and scribes; others as hunter-gatherers.[1] To Francisco Pizarro, the Inca Empire was a transfer of resource frontier. To the instigators of the rubber boom in western Amazonia at the end of the 19th century, the Zaparo territory was a transfer of resource frontier—and once the Zaparos were thus weakened, their longtime competitors, the Huaorani, moved in as well.[2] In modern times, transfer of resource frontiers are everywhere: oil drilling, fracking, and logging in ancestral lands; predatory lending, as with subprime mortgages and much student debt; the Holocaust. One symptom of transfer of resource frontiers is tyranny.

Geographic frontiers represent the discovery of resources heretofore unknown to humans. Geographic frontiers are inherently zero-sum: there is a finite amount of space on this planet of ours, and we will reach the end of it. Technological frontiers are the creation of resource through human ingenuity. Technological frontiers are temporarily non-zero-sum—specifically, positive-sum—and this can appear to be a permanent state.

But there are physical limits: a single electron is the theoretical minimum needed to flip from one state to another in a transistor, for instance. Transfer of resource frontiers are theft of resource from other human populations. Like all frontiers, transfer of resource frontiers are ultimately zero-sum. Theft has its limits; even thieves must obey physical laws.

What choice do we have but to continue seeking new frontiers, and more growth? If our addiction is just a special human case of a pattern that characterizes all species that have ever lived, are we not simply condemned to ride this destructive trajectory to the end?

We have written this book, in part, because we believe the answer to that question is *no*.

Humans are obsessed with growth because it engenders bigger populations that, if nothing else, have farther to fall before going extinct. But large populations also pose a hazard to themselves if the resources that enlarge them are finite or fragile. In such cases, moderation is the key, but it only works if our drive for growth, our individual perception of it, is sustainably satisfied.

We have run out of geographic frontiers, or nearly so. Technological frontiers—by turns dazzling and disappointing—come with risks (beware Chesterton's fence!), and are ultimately constrained by available resources. Transfer of resource frontiers are immoral and destabilizing. What, then, are we to do? Where to turn to find salvation? In simple terms, consciousness. Consciousness can point the way to a fourth frontier.

Once again: The human niche is niche switching, and consciousness is the answer to novelty. Living sustainably on a finite planet is a hard sell, but we can, and we must, find a way. We have no choice. These problems of novelty require humanity's urgent attention, and cannot be solved by the goodwill or hard work of individuals.

We moderns have become a threat to our own persistence. We are built to figure out how to move between modes of being. It is time to rise to collective consciousness and prototype a way out of this.

We face some significant obstacles. Humans, like other creatures, are obsessed with growth, and we are capable of driving ourselves extinct in the pursuit of it. Even though it is logically obvious that we must accept equilibrium, we are not built to be satisfied with it because being unsatisfied has been an excellent strategy for the last several billion years.

There is a character trait of individuals that may be critical to finding the fourth frontier, or rather, its adaptive foothill that could lead to a society-wide solution. That is: pride in craftsmanship. An artisan who takes pride in the quality and durability of their work is enacting some portion of a fourth frontier mentality, one in which the life span of a product is as important as its function. A table or sideboard made by a local craftsperson is not beloved merely because it is more beautiful than what can be assembled from a box bought at Ikea, but also because the person in possession of a lovely and functional piece has a chance of handing it down to their children, or other kin, or friends. So, too, would we like to be able to deliver unto the next generations a lovely and functional world.

The fourth frontier is a framework, therefore, that can be understood with an evolutionary tool kit. It is not a policy proposal. The fourth frontier is the idea that we can engineer an indefinite steady state that will feel to people like they live in a period of perpetual growth, but will abide by the laws of physics and game theory that govern our universe. Think of it like the climate control that allows the inside of your house to hover at a pleasant spring temperature as the world outside moves between unpleasant extremes. Engineering an indefinite steady state for humanity will not be easy, but it is imperative.

Senescence of Civilization

We are headed for collapse. Civilization is becoming incoherent around us. In organisms, we know what causes senescence (the tendency to grow feeble with age). It is antagonistic pleiotropy, the propensity of selection to favor heritable traits that provide early life benefits even when they carry inevitable late life costs.[3] This willingness to accept harm in old age occurs because selection sees the early life benefits much more clearly, as individuals will often reproduce and die before the harms have time to fully manifest.

There is an analogous argument to be made for the senescence of civilization. Our economic and political system, in combination with our desire for growth in the moment, inflicts policies and behaviors that don't seem crazy at first, not at all, and yet they too often turn out to be not only bad for us and the planet, but also irreversible, by the time we realize what we have wrought. We are living the unfortunate reality of the Sucker's Folly—again, the tendency of concentrated short-term benefit to not only obscure risk and long-term cost, but also to drive acceptance even when the net analysis is negative.

When dimensional lumber began to be produced, it surely seemed like a pure boon; who could have predicted that living in a world of carpentered corners would literally change how we see? The first time that someone put a distillate of oil in a motor and made it run, you would have seemed crazy to say that you shouldn't do it. Even things that appear to be an unalloyed good generally have risks. Being able to listen to music without disturbing others was a breakthrough. As we now understand, though, it is easy to listen to music through headphones—or worse, with earbuds— at volumes that are destructive to our hearing. What we "want," and what the market is glad to hand us, is short-term gratification that rarely accounts for what is best for us long term. A market that is unregulated will tend to embody the naturalistic fallacy—the mistaken idea that "what is" in nature is "what ought" to be. When we let such unregulated markets lead, we are fed directly into the naturalistic fallacy. Just because you can doesn't mean that you should.

Compounding the issue of unregulated markets is the reality that humans are perfectly adapted to manipulate one another, and that such adaptations have been moved into the hyper-novel territory of widespread anonymity. Historically, manipulation was kept in check by living in small groups of interdependent people. Shared fate was the rule that kept us in line. Putting one over on a person whose fate is intimately linked to your own is generally a poor idea, and those who do quickly get a reputation for doing so. We no longer live in small, interdependent communities. Many of the most critical systems we rely on are global, and the participants are nearly always anonymous. Malicious market forces are largely an expression of manipulation made possible by such anonymity, and by a lost sense of shared fate.

With all this stacked against us, how do we move forward? Civilization as we know it is going to become senescent because that which made us successful will ultimately destroy us. The answer, in simple terms, is to consciously build a system that is resistant to senescence. Doing so is much more complicated, but here are a few ideas to get us started.

The key to building a system that is resistant to senescence is to:

- Not optimize for a single value. Mathematically speaking, if you try to optimize for any single value, no matter how honorable—be it liberty or justice, decreasing homelessness or improving educational opportunities—all other values, every single other parameter, will collapse. Maximize justice, and people will starve. Everyone may starve equally, but that's small recompense.
- Create a prototype for your system. After that, continue to build prototypes. Do not imagine that you know from the beginning what the final system will look like.
- Recognize that the fourth frontier is inherently a steady state, whose characteristics are ours to define. We ought to strive to create a system that:
  - Liberates (that is, that frees people to do rewarding, interesting, awesome stuff),
  - Is antifragile,
  - Is resistant to capture, and
  - Is incapable of evolving into something that betrays its own core values. In the technical language of evolution, we need a system that is an Evolutionarily Stable Strategy, a strategy incapable of invasion by competitors.

The Maya

In many regards, the problems we are facing now are ones we have faced before. Every culture in human history has engaged in both cooperation and competition, and behaved both in ways that ought to make us proud to be human and in those that make us ashamed. Both glorious and ghastly actions have been widespread.

When looking back on history, we have the responsibility to recognize that truth, and also to recognize when our ancestors' wins—legitimate or, very often, not—have provided us advantage that we did not ourselves earn. It is not, however, our responsibility to subjugate ourselves to those histories.

Europeans indeed stole land from Native Americans, in often gruesome and despicable ways. The Native Americans who were thus subjugated themselves had a history of warfare and conquest in the New World, taking land from one another. And, of course, none of this was new—they brought it with them to the New World when they crossed over from Beringia many thousands of years earlier.

Let us not romanticize any people or period. Let us instead understand humanity holistically, and work to provide opportunity equally to everyone going forward.

In this book, we have shared an evolutionary tool kit with which to understand the human condition, not to justify it. We are not served by ignoring what we are—brutal apes, by one measure. We are also not served by pretending that brutal apes are the only thing that we are. We are also generous, cooperative beings full of love. We have arrived in the 21st century with evolutionary baggage, and a fair bit of intellectual confusion. Let us understand the baggage, in order to reduce the confusion, and increase our odds of moving forward with maximal human flourishing.

As an aid to this end, let us consider the Maya.

The Maya thrived for over two and a half millennia in Mesoamerica, surviving droughts and enemies and other unpleasant extremes. In the now ancient city-states of the Maya—including not just Tikal, but also Ek' Balam, Chacchoben, and so many more—stone pyramids and temples are still visible above the tops of the trees. On the forest floor, footpaths run between ancient buildings, as do agoutis, lizards, and the occasional ocelot. More substantial roads, *sacbes*, connect city-states. Most of the Mayan

city-states emerged as political, economic, and cultural forces to be reck-
oned with long before the Roman Empire existed. Wholly unaware of the
other's existence, the Maya and the Romans were at their peak at the same
time, in the early part of the first millennium, and both were in obvious
decline by the beginning of the second.

The Maya had an Enlightenment of their own, long before the European Enlightenment. We will never know its extent, as the vast majority of
their books were destroyed by Europeans.

The Mayan civilization was spread widely across the Yucatán peninsula, extending south through modern Belize and Guatemala, and just
barely dipping into Honduras. The Maya were dominant in these landscapes for twenty-five hundred years, but they were not monolithic, and
their successes waxed and waned over both time and space. City-states
collapsed, droughts caused the abandonment of once-fertile lands, and
while some areas were repopulated by the Maya, others never were.[4]

The Maya were intensive agriculturalists who farmed on poor tropical
soils but managed to maintain soil fertility for a remarkably long time
through successful land management. They dealt with the hilly slopes that
were ubiquitous through much of their range with at least six types of
terracing systems. They used complex reservoirs to conserve water during
annual dry seasons, and during less predictable, longer dry spells. It is also
true, however, that where they cleared forests, the land was generally de-
graded, and soil quality fell.[5]

By the time the Spaniards arrived, the Maya were already in decline.
They had had a long run, and what, exactly, precipitated their collapse is
up for debate. While the Mayan culture largely disappeared, the Mayan
people persist. They were not a fragile people or culture. They were robust and long-lasting. One indication of just how long a run they had is
that they had a unit of time, the *baktun*, equal to 144,000 days—almost
four hundred years. They were so long-lived, as a people, and so accustomed to thinking across long time spans, that they used the baktun to
help keep track of time.

The durability of the Maya suggests that the potential exists for conscious, directed enlightenment, in which we take ownership of our own evolutionary state. Like the Maya, we moderns need to find ways to flatten the boom-bust cycle that has plagued all populations across time. We hypothesize that the Maya did this by creating a mechanism for not turning excess resources into more people, or ephemeral things; instead, they invested in giant public works projects. Many of these public works projects are visible today as temples, as pyramids. They grew them like onions, building more layers in times of abundance. In years of plenty, we posit, when excess food could easily have been turned into more people, which would have expanded the population, making hunger and conflict inevitable in lean years, the Maya instead turned the extra food into pyramids, or into bigger pyramids. They created glorious and useful public spaces, enjoyable by all, and when agricultural boom years inevitably ceded to bust years, the temples required no nourishment, and the population could withstand the leaner times.

Western civilization has been dominant for nearly as long as the Maya were. Their culture unraveled, accelerated at the end by a hostile enemy from across an ocean. Our culture is unraveling as well. We need a new steady state, an Evolutionarily Stable Strategy. We need to find the fourth frontier.

Obstacles to the Fourth Frontier

Many forces are obstacles to the fourth frontier. Trade-offs persist even once they are recognized; obsession with growth blocks progress that doesn't look or sound like growth; and regulation is difficult to get right. None of these obstacles are insurmountable, but they are large. In the next three sections, we address each in turn.

TRADE-OFFS IN SOCIETY

Just as no bird can be both the swiftest and the most agile, no society can be both the freest and the most just. Freedom and justice exist in trade-off relationship with each other. We ought not try to push either of these two sliders all the way to one end.

It is of course true that many societies are both less free and less just than they might be. For a majority of situations, we are not yet up against the limit of what is possible (what economists call the efficient frontier), and could potentially increase both freedom and justice until that limit is reached. Coming to grips with the fact that freedom and justice cannot both be maximized is a critical step in the conversation, though. Imagining a world that is totally free and just is to imagine utopia, a static perfection, a world in which trade-offs have been banished. Utopia is an impossibility, and its persistence as a fantasy is a profound hazard.

Within a democracy, one way—but hardly the only way—to divide the political feelings of the populace is liberal versus conservative. Left versus right. Liberals and conservatives tend to have distinct blind spots, particular ways of misunderstanding or conveniently forgetting about trade-offs. While we are writing with American terminology, these observations hold across national borders.

In order for us to have a conversation about humanity's future effectively, people of every political persuasion need to understand diminishing returns, unintended consequences, negative externalities, and the finite nature of resources. Liberals (our political kin) are particularly prone to underestimating *diminishing returns* and *unintended consequences*. Conservatives are particularly prone to underestimating *negative externalities* and the *finite nature of resources*.

According to the economic law of *diminishing returns,* as you increase your input to a given variable, while holding everything else constant, the increases in your yield will virtually grind to a halt. Diminishing returns occur in every complex adaptive system. Understanding this encourages us to create nimble, evolving strategies, rather than cumbersome, static ones. A utopian vision, one that seeks to maximize any single parameter, falls prey to diminishing returns. Because we are constantly reaching for a static goal—which requires greater and greater investment to achieve, with ever smaller gains—we greatly limit what else could be accomplished.

Diminishing Return Curves

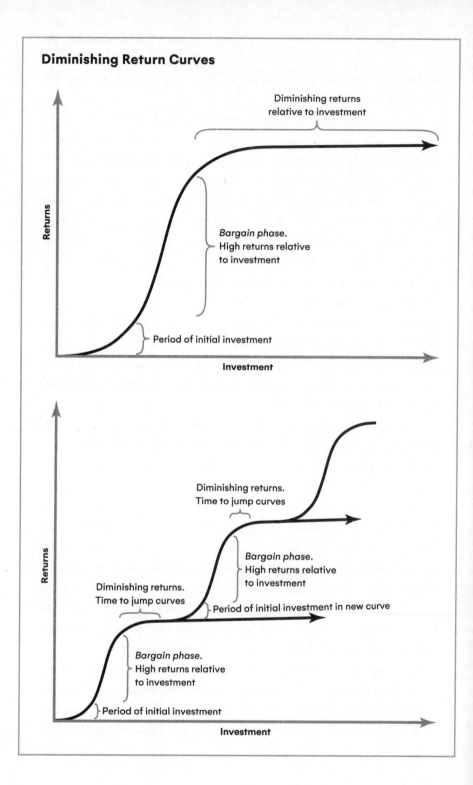

The opportunity cost of not jumping to the next diminishing returns curve is spectacular.

Unintended consequences are a variant of Chesterton's fence: messing with an ancient system that you do not fully understand may create problems that you do not foresee. Liberals are prone to making regulations that upset functional systems. For instance, linking education funding to test scores led to the unintended consequence of creating a feedback loop in which poor scores reduce funding, which then further decreases scores. That said, conservatives are prone to easing regulations to facilitate the creation of new products—which themselves may upset functional systems. For instance, deregulating waste management to reduce operating costs has created pollution, which is effectively externalizing the costs of waste management. This has destabilized uncountable natural systems that humans historically depended on: fish and shellfish that are too toxic to eat, rivers that can't sustain fish populations, air quality that leads to asthma and developmental delays. In short, both liberal solution making and conservative desire for market innovation are the source of unintended consequences.

Negative externalities occur when individuals making decisions—or products—do not have to bear the full cost of those decisions. Consider Ankarana, a remote and glorious nature reserve in far northern Madagascar. It is a 150-million-year-old limestone plateau whose roof of razor-sharp ridges has collapsed in places, creating a network of caves through which underground rivers run. These caves open into wholly isolated pockets of forest, filled with crowned lemurs and geckos. It is a landscape and biota unlike any place else on Earth. Alas for Ankarana and its inhabitants, it is also home to vast sapphire deposits that, when we were there in the early 1990s, were being mined for jewelry and industrial grit, to the obvious detriment of the environment. Wherever those extracted sapphires ended up, it is certain that few if any of those who benefited from the stones knew anything of the damage their extraction had wrought. This is a negative externality, which can propagate because money is fungible, allowing the *harm* from what is created to be disassociated from its *value*. Ankarana is one easily understood example, but negative externalities are everywhere. From the burning of coal for energy, in which the air pollution is shared by all but the profits by few, to the playing of loud music late at night to the enduring irritation of your neighbors, negative externalities are rampant in our world.

The *finite nature of resources* should be obvious. While there are some resources that are effectively infinite—oxygen and sunlight being at the top of that list—the vast majority of Earth's resources are finite. From rubber to wood to oil, from copper to lithium to sapphires, all are limited.

The partisan nature of Western democracy can make us feel as though we could never align behind a set of shared values, but realizing that we have a lot in common is the only way to achieve collective consciousness. We have but one planet. And yet we continue to behave as though the world in which we live is a cornucopia of infinite wealth. The Sucker's Folly blinds us, our nature seeks growth, and our culture, lagging behind the times, is wired for a world in which we no longer live. While the Omega principle reveals that our culture is not arbitrary, it does not guarantee that our culture is up to snuff when it comes to hyper-novelty. This is the domain of consciousness.

OBSESSION WITH GROWTH

The American Dream was a fiction, but it wasn't wholly a fiction. It had elements of the fourth frontier, but it was also based on a utopian fantasy of infinite growth. One big cultural battle we're in now is that while some of us understand that infinite growth cannot last, others among us are cornucopians.

The evolutionary creature in all of us needs to feel growth. Growth is what winning feels like, in evolutionary terms. Every single one of us, every lineage that has existed on Earth, has been through an oscillating cycle of growth, of filling up a niche, and of running into the end of the resource—of moving from a non-zero to a zero-sum world.[6] Coming up short against that limit feels terrible, whereas abundance allows humans to flourish.

Chasing growth as if it is always there to be caught is a fool's errand. Sometimes the opportunity exists, and sometimes it doesn't. The expectation of perpetual growth is in many ways similar to the pursuit of perpetual happiness—it is the route to a host of miseries.

Our obsession with growth and the economic mindset it has created has given rise to a throughput society: one in which civilization's health is evaluated based on the production of goods and services, where more consumption is presumed to be better. This framework is so deeply embedded in our minds that it seems almost logical, until one considers the implications.

Imagine the introduction of a new type of refrigerator, one that lasts much longer than other models, is similar in cost, and is equal to them in performance. A healthy society would regard this as a good thing, as would most citizens, reducing waste and pollution, conserving energy and materials, and possibly limiting strategic vulnerabilities that come from heavy dependence on foreign suppliers. But the effect of this durable fridge on gross domestic product will be negative—and so indicative of a problem. Now imagine that we achieved similar gains in durability in all consumer goods. With goods needing less frequent replacement, we'd face a massive economic contraction. Jobs would be lost, incomes would drop, taxes revenue would decline. In short, it would destroy the ability of our system to function.

Similar absurdities emerge anywhere that something positive interrupts demand. Would it be good if people invested more time and effort with their romantic partners instead of paying for porn? Would it be good if people were more satisfied with what they had and less susceptible to sales pitches? Would it be good if people were more easily sated and less prone to overeat? Would it be good if people spent more time producing art, music, and insight and less time coveting, purchasing, and flaunting trendy goods? Of course. All of these things would be major upgrades to our way of life. But our growth-obsessed economic mindset would report exactly the opposite. Our throughput society depends on insecurity, gluttony, and planned obsolescence. It's how we keep the lights on.

Our obsession with growth has thus been a mixed bag. It brought us this far, at the cost of a great deal of suffering and misery. Nevertheless, with more than seven billion people on the planet, consumption cannot remain our measure of well-being. If we are to persist, sustainability must displace growth as the indicator of success.

When we were in the Trinity Alps of far northern California in the summer of 2019, the lack of animals was notable. On a three-hour hike we saw only a handful of birds. Summer road trips no longer produce windshields smeared with the carcasses of insects; roadkill is becoming less common. In early 2020, when we were in Yasuní—Ecuador's jewel of a national park in the western Amazon, understood to be the most biodiverse place on the planet—there were fewer insects than there used to be, and fewer birds.[7] We suspect that among other culprits, the die-off of birds and insects is due to widespread use of insecticides, far upstream of (or more distant yet from) the Amazon. Insecticides get aerosolized and fall into the water, which then comes downstream, out of the Andes. Once the insects are gone, so too go the insectivorous birds and bats and lizards, and once they're gone, so too go the carnivores—the tayras and short-eared dogs and jaguars. Rachel Carson was right—but the silent spring of the northern temperate zone has arrived in the tropics, and it is a harbinger of greater dangers to come.

Some will look at an analysis like this and say, "Sure there are problems, but people are always predicting the end of the world, and they haven't been right yet." But that isn't the correct way to think about it.

The "world ending" doesn't typically mean the destruction of the planet. Rather, it means something like "our world"—in effect, our ability to persist into the future. When it is framed that way, some individuals who have predicted the end of their world will certainly have been right. Many populations have, after all, faced threats to their survival, and many of those failed to rise to the challenge. It is our belief, therefore, that sensitivity to existential threats is a long-standing adaptive trait, and that the size of the present human population, our degree of interconnection, and the technology we now possess all create a threat to our species analogous to threats faced by ancestral populations. The problem is old, the scale brand-new.

ON REGULATION

Good laws and regulations are hard to write. Simple, static laws will either be wrong from the beginning or have a short shelf life. Having a short shelf life is fine, to the extent that the system can upgrade. As Thomas Jefferson observed, even democracies need rebellion with some regularity.[8] To the degree that a system is set in stone, it will be both gameable and gamed.

Evolved systems that have persisted over time are generally complex and functional, and we should employ the Precautionary principle when tinkering with them. Removing functional organs because we can't tell what they're for is not wise. It's tempting therefore to laugh at those doctors who once proposed to take out people's healthy large intestines, but what similar mistakes are we making now? Given the hyper-novelty of our era, it would be the height of arrogance to imagine that we are not currently doing things that will be understood to be laughable, even deranged, in the future.

Society is obsessed with short-term safety because short-term harm is easy to detect and comparatively simple to regulate. Long-term harms are a different story, being more difficult to detect, and even harder to prove. What *are* the long-term effects of screen time or educational testing, aspartame or neonicotinoid insecticides? We do not know. But because no one wants to live in a world where safety testing keeps every innovation off the market for decades, we have become reckless. We foolishly presume long-term harms are absent until they can no longer be ignored, and are then shocked that our expectations of safety were wrong.

Regulation has a bad reputation in many circles. It is often done badly, and when it is done well, it tends to render the problems it addresses minor or invisible. As such, many see it as a needless obstacle, unaware of the benefits it has brought. A good regulatory scheme is efficient and light-handed—all but invisible. While inherently constraining, its net effect should be liberating, allowing access to the benefits of innovation without having to obsess about hidden consequences.

Good regulation is a key ingredient in any functional complex system. Our bodies, for instance, are tightly regulated across many domains, including temperature. To keep us within optimal range, myriad systems constantly adjust the balance between heat generated and lost, shunting blood into and away from our extremities and capillary beds. Our temperature is nothing if not well regulated, but these processes are so effective that they are barely detectable to us, freeing us to do anything from swimming in a cool river to playing soccer in the sun, all while rarely giving a thought to the risk of hypothermia or heatstroke.

No manufactured system has regulation that is as elegant as that found in the human body, but we do have examples of good regulation in such systems. Take commercial air travel, perhaps the safest way to travel. Its safety is due to every aspect of it being regulated, and systematic investigation of the rare accidents that do happen. One could complain about the cost and inefficiency of the rules surrounding aviation, but those objections must be understood in context: the regulations make it possible for a sizable fraction of the world's population to access almost any location on Earth within twenty-four hours, all while being far safer than one is when driving to the airport. The cost of the regulations is small compared to the freedom they create, a goal we should strive for with every industrial process.

Large systems outside the scope of possible containment by individuals need to be regulated. We cannot address nuclear safety or oil extraction or habitat loss without large-scale regulation.

Leveling Up

We need as many people as possible to get on board with this discussion, to mature into adults, and to discard their utopianism. We need people to welcome the idea that some set of values has to be broadly embraced and pursued, and to recognize that we're not going to arrive at a decent future by describing it precisely in advance. We're going to reach it by agreeing on the characteristics that such a desirable and plausible world must have, and then we must prototype, evaluate the results, and prototype again. We must find the foothill, and discover the path from there. We will navigate through the fog, and we will not have a blueprint. We must start now. We can't wait until the hazard is so clear that we all agree—then it will be too late.

We are in the throes of a sustainability crisis. One thing or another will take us out. It might be climate change, or a Carrington Event, or a nuclear exchange set in motion by wealth inequality, a refugee crisis, or revolution, to name just a few of the awfully real possibilities. We are hurtling toward destruction. We must, therefore, with full consciousness, embark on something dangerous. We must seek the next frontier: the event horizon, beyond which we cannot see, from which we cannot return, but through which may be our salvation.

The Beringians could not detect that the New World existed, but they could not stay in the Old World. They headed east into the unknown, into a forbidding landscape of rock and ice, of rough seas and dangerous terrain . . . and ultimately, into two vast continents of plenty.

Polynesians left their ancestral homes, crossed vast open ocean, and many of them must have died, but some of them discovered and colonized Hawaii. Others, moving west across the Indian Ocean rather than east across the Pacific, discovered and colonized Madagascar.

People have been discovering new worlds for as long as we have been human, but we are now out of geographic frontiers. Now we must discover a new world again, and become emergent. We must seek the foothills of peaks that are higher and more promising than the ones that we are now on. We must become more than our best selves, and save ourselves in the process.

The Corrective Lens

- **Learn to hack and kludge your own mental architecture for a better life.** Keep markets as far from your motivational structure as you can—don't let someone else's profit motive determine what you desire or do.
- **Keep commerce away from children, for as long as possible.** Children raised to put high value on the transactional nature of being become dedicated consumers. Consumers are less observant, meditative, and deeply thoughtful than people who value creating, discovering, healing, producing, experiencing, communicating.
- **Individuals need to calm down, and level up.** Rely less on metrics, more on experience, hypothesis, and deriving truth and meaning from first principles. Rely less on static rules, and seek an understanding of the context in which those rules are appropriate.
- **Dispense with anything predicated on a utopian vision that focuses on a single value.**
  - As soon as someone reveals that they are trying to maximize a single value (e.g., freedom or justice), you know they are not an adult.
  - Liberty is emergent—therefore not a single value. It is an emergent consequence of having fixed the other problems (e.g., justice, security, innovation, stability, community/camaraderie).

Society-wide, we should:

- Like the Maya, invest our surplus in public works, which make us antifragile.
- Prototype, prototype, prototype.
- Move to a precautionary mindset, such that we can learn to regulate our industries effectively, minimizing any negative externalities that they create.
- Consider Chesterton's fence in all of its guises—from health care to cuisine, from play to religion.

From the moment when our ancestors achieved ecological dominance, competition between populations has been our dominant selective force.[9] Millions of years of evolution have refined our circuitry for such competition, and it has become the default at the human software level. Now, though, three things conspire to make the inclinations that brought us to this moment an existential threat to our future: the scale of the human population; the unprecedented power of tools at our disposal; and the interconnectedness of the systems on which we depend (global economy, ecology, and reach of technology).

The importance of understanding human software is urgent. The problem we face is the product of evolutionary dynamics. All plausible solutions involve awareness of those dynamics.

The problem is evolutionary. So is the solution.

Tradition, and How to Tweak It

In our home, one of our annual rituals is to celebrate Hanukkah, the Jewish festival of lights that occurs just before or around the northern winter solstice. We light the menorah, as is traditional, and each night review an additional principle, which is not.

Our family's new Hanukkah rules:

- Day 1: All human enterprises should be both sustainable and reversible.
- Day 2: The Golden Rule: Do unto others as you would have them do unto you.
- Day 3: Only support systems that tend to enrich people who have contributed positively to the world.
- Day 4: Don't game honorable systems.
- Day 5: One should have a healthy skepticism of ancient wisdom, and engage novel problems consciously, explicitly, and with robust reasoning.
- Day 6: Opportunity must not be allowed to concentrate within lineages.
- Day 7: Precautionary principle: When the costs of an action are unknown, proceed with caution before making change.
- Day 8: Society has the right to require things of all people, but it has natural obligations to them in return.

Afterword

IN JANUARY OF 2020 WE WENT TO THE TIPUTINI BIODIVERSITY STATION
in the Ecuadoran Amazon to finish our first draft of this book. When we
emerged from our isolation—as our phones came alive for the first time in
two weeks—we were confronted with a barrage of news, mostly trivial, of
which we had been blissfully unaware. But in that onslaught there was one
ominous report—a case of a "novel coronavirus" in Ecuador. The pathogen
came from horseshoe bats, had jumped to people and then spread rapidly,
first in Wuhan, China, and then beyond.

As the two of us tried to make sense of these first hints of pandemic,
it quickly became clear that there might be more to the story. Wuhan, we
soon learned, housed a BSL-4 laboratory—it was, in fact, one of our planet's two main centers of research on bat-borne coronaviruses. These viruses were being studied in Wuhan and in North Carolina because of fear
among scientists that such viruses *could* jump to people and, without very
much evolutionary change, cause a dangerous pandemic. If nothing else,
the fact of the pandemic having begun in one of two cities where these
viruses had been under intensive study seemed a spectacular coincidence.

As of the writing of this Note in late May, 2021, the consensus in the
scientific establishment, including national and international regulators,
and in the mainstream press that follows them, has finally shifted to one
of grudging acceptance of the obvious: SARS-CoV-2 may well have leaked
from the Wuhan Institute of Virology, and the COVID-19 pandemic might
therefore be for humanity an entirely self-inflicted wound. The strength of
this hypothesis is something we have been discussing on our podcast,

DarkHorse, since April of 2020. Those discussions caused a great deal of derision and stigma to be directed at us, and it is a bewildering relief to watch the world suddenly come around to the plausibility of this well supported, if unfortunate, explanation.

But no matter what humanity ultimately concludes about this pandemic's origin, there is a deeper truth hovering just outside our collective awareness: COVID-19 is a product of technology, no matter what path it took to humans.

Consider this fact: from the beginning of the pandemic the virus showed essentially zero capacity to transmit outside. Put another way, COVID-19 is a disease of buildings, cars, ships, trains, and airplanes. More than 99 percent of the Earth's surface is a COVID-safe zone. Even in your own backyard the virus will struggle mightily to infect anyone—it has no meaningful impact unless you caught it before you walked out. In the park, on the balcony, at the beach, we are—at least for now—immune.

The dependence of the virus on enclosed spaces also means that, had humanity agreed to avoid these vectoring environments for a few weeks, the pandemic could have been quickly brought to a halt. But this scenario in which we free *ourselves*, and lock down *dangerous environments* instead, is little more than an idle thought experiment. Even though, in evolutionary terms, these dangerous environments are all brand new to humans, the idea of humanity staying outside of them, even for just a few weeks, is unthinkable.

Many individuals could do it, but the majority would be at a total loss, even though we evolved outside, and despite the fact that most of our ancestors would have spent every hour of their lives in what we now strangely call "the outdoors." We have forgotten the skills we once knew so well. That knowledge of, and comfort with, our natural environment has been replaced with a different skill set, one tuned to pursuing value and avoiding harm in a synthetic environment of our own device. Our cognitive software has been rewritten, and we have forgotten too much to ever again be what we were. As a result, we are condemned to battle this pathogen in bespoke environments on which we and it have both grown to depend.

That's the view on the ground, but the human dimension of this pandemic is even clearer from 30,000 feet. Or more accurately, *at* 30,000 feet. For it is the way we have begun to travel that really set us up for pathogenic disaster. SARS-CoV-2 crossed oceans in hours, and it didn't pioneer some ingenious new mode. Where once an epidemic might have been held back by barriers that limit human travel, humans now regularly transmit communicable diseases from their continents of origin to every corner of the globe.

Much as people thought little about washing their hands prior to the germ theory of disease, we give no thought to the scale of misery caused by a given person transporting a new and nameless cold virus to some continent that was free of it the day before. "Novel Coronavirus" took advantage of that nonchalance before the pathogen even had a proper name.

The COVID-19 pandemic is itself a symptom of another disease entirely. In the pages of this book, we call that disease "hyper-novelty." It is caused by a rate of technological change so rapid that transitions in our environment outstrip our capacity to adapt.

You will not find the COVID-19 pandemic specifically dissected here, but you will find a full exploration of the hyper-novelty crisis that left us vulnerable to this virus—a virus so weak that it could have been cured with a bit of well-coordinated fresh air.

Acknowledgments

We stand on the shoulders of giants. Of those whom we have known and learned from personally, Richard Alexander, Arnold Kluge, Gerry Smith, Barbara Smuts, and Bob Trivers loom particularly large. We knew Bill Hamilton and George Williams less well, but their influence on us was profound, as was that of many of our contemporaries, including Debbie Ciszek and David Lahti. Early conversations between Bret, Jordan Hall, and Jim Rutt, in which they imagined an alternative to our current, broken paradigms, became known as Game~B, of which the Fourth Frontier is a variant. Later, with Mike Brown at his "Science Camp" on Double Island, a few of us would continue those conversations.

For further development of our ideas, we wish to thank our students at The Evergreen State College, to whom we delivered some of the thinking in this book. In particular, the students of Adaptation, Animal Behavior, Animal Behavior and Zoology, Development and Evolution, Evolution and Ecology Across Latitudes, Evolution and the Human Condition, Evolutionary Ecology, Extraordinary Science of Everyday Experience, Hacking Human Nature, and Vertebrate Evolution provided wit, challenge, and insight as we played with and developed concepts and connections.

Among those many excellent students is Drew Schneidler, who was also our research assistant on this book, and who has long been a friend. We first came to know Drew in 2007, and he would later be with Heather on the first study abroad program she created at Evergreen. His brilliance across domains helped shape this book, and he was truly our collaborator. Time and again, Drew cut the Gordian knot when it seemed uncuttable.

We thank, as well, our early readers, who gave generously of their time, effort, and skill: Zowie Aleshire, Holly M., and Steven Wojcikiewicz.

As our academic lives at Evergreen splintered into a thousand shards in 2017, we were lucky to have families who never wavered in their support. Countless more people showed up and helped us as well; without them, this book is unlikely to have emerged in its current form. Within the college, these people included but were not limited to Benjamin Boyce, Stacey Brown, Odette Finn, Andrea Gullickson, Kirstin Humason, Donald Morisato, Diane Nelsen, Mike Paros, Peter Robinson, Andrea Seabert, and Michael Zimmerman. Outside of Evergreen, just a very few of these are Nicholas Christakis, Jerry Coyne, Jonathan Haidt, Sam Harris, Glenn Loury, Michael Moynihan, Pamela Paresky, Joe Rogan, Dave Rubin, Robert Sapolsky, Christina Hoff Sommers, Bari Weiss, and Bob Woodson. We are also grateful to Jordan Peterson, for blazing a trail that helped us find our way through the darkest times at Evergreen, and for modeling intellectual integrity under fire.

The most unflinching and fearless among these many intellectual and political influences and allies has for a very long time been Bret's brother, Eric Weinstein.

We also want to give particular thanks to Robby George, and the James Madison Program in American Ideals and Institutions at Princeton University, for temporarily lifting us out of academic exile, and welcoming us as visiting fellows while we wrote this book.

Our agent, Howard Yoon, of Yoon Ross Agency, was yet another person who reached out to us as Evergreen was blowing apart. To our relief, he was not interested in "the Evergreen book." We discussed several projects before collectively realizing that this—a bit of everything, all framed evolutionarily—was the right one, and indeed, one that we had been talking about writing for many years. We were just finishing the proposal when Helen Healey, now our editor at Portfolio/Penguin, made first contact. Both Howard and Helen have been staunch supporters and valuable sounding boards throughout the process.

The Tiputini Biodiversity Station, in the Ecuadorian Amazon, provided respite and insight for a few brief weeks as we were finishing the first draft of the book. Kelly Swing, who is both the founding director of Tiputini and our friend, along with the excellent staff, are working hard to preserve wild nature in one of the most remote outposts in the world. It feels imperative that they be successful.

Finally, we are grateful to our children, Zack and Toby, who are seventeen and fifteen years old, respectively, at the time of publication. They grew up exploring landscapes from the Pacific Northwest to the Amazon with us, first privy to, then contributing to, many of the conversations that became this book. We never wanted them exposed to the losses and realities of modern humanity that the blow-up at Evergreen revealed to all of us, but they have shined. We are lucky to have such remarkable young men in our lives.

Glossary

Some of these definitions have been borrowed in part or in whole from Lincoln, R. J., Boxshall, G., and Clark, P., 1998. *A Dictionary of Ecology, Evolution and Systematics,* 2nd ed. Cambridge: Cambridge University Press.

adaptation: The process by which selection on heritable traits (*sensu lato*) increases capacity to exploit an opportunity.

adaptive landscape: A metaphorical framework used in conceptualizing how selection and adaptation operate. Introduced by Sewall Wright in 1932,[1] a brief explanation can be found in note 19 for chapter 3.

alloparenting: Caregiving behavior, like that of a parent, provided by an adult toward an individual who is not their direct offspring.

antagonistic pleiotropy: A form of pleiotropy (wherein one gene has effects on multiple traits) in which the fitness effects run counter to one another. Regarding senescence, one effect is beneficial early in life, while another is detrimental late in life.

antifragile: The state of increasing in capability when exposed to stressors or harm, as coined by Nassim Taleb in 2012.[2]

Beringia: A landmass that emerges from the Bering Strait during ice ages, when sea levels drop. Likely an ancestral habitat for all sub-Arctic natives of the New World.

carrying capacity: The maximum number of individuals that can be stably supported, at equilibrium, by a given spatiotemporal opportunity (e.g., X number of wolves in Yellowstone in 1900).

Chesterton's fence: The idea that reforms should not be made to a system until the reasoning behind its current state is understood. Originally described by G. K. Chesterton in 1929.[3]

clade: An ancestral species and all of its descendants. Synonymous with a monophyletic group; ideally synonymous with taxon. Examples include birds, mammals, vertebrates, primates, and whales (which includes dolphins).

conspecific: A member of the same species.

consciousness (when contrasted with culture, for the purposes of the model in this book): The fraction of cognition that is packaged for exchange between individuals (e.g., thoughts that can be communicated).

culture (when contrasted with consciousness, for the purposes of the model in this book): A package of adaptive beliefs and behavioral patterns transmitted outside of the genome. Most culture is transmitted vertically; culture is different from genes, however, in that it may also be passed horizontally.

Darwinian: The tendency to adapt in response to differential success of heritable traits. Referencing Charles Darwin, who first identified both natural and sexual selection.

Environment of Evolutionary Adaptedness (EEA): The environment that favored the evolution of a given adaptive trait. Humans have many EEAs, not just that of the African savannahs and coasts inhabited by our early hunter-gatherer ancestors.

epigenetics:

> *sensu stricto*: Regulators of gene expression that are not encoded in the DNA sequence itself (e.g., DNA methylation).

> *sensu lato*: Any heritable trait not due directly to changes in DNA sequence. It includes epigenetic (*sensu stricto*) phenomena and, for example, culture.

eusociality: A social system in which some individuals forgo reproduction in order to facilitate the reproduction of others. Eusocial populations function as superorganisms, with shared interests and shared fate.

Evolutionarily Stable Strategy: A tactic that, once adopted by most members of a population, is invulnerable to displacement by competing strategies.

first principles: The most fundamental and secure assumptions relative to a domain (akin to axioms in math).

frontier (for the purposes of the model in this book): A non-zero-sum opportunity for a population. Three established types: geographical, technological, interpopulation transfer.

gamete: A mature reproductive cell that fuses with another to form a zygote.

game theory: The study and modeling of strategic interactions between two or more individuals. Particularly salient when the optimal strategy depends on the moves most likely to be adopted by others.

generalist: A species or an individual with broad tolerances, or adapted to a very broad niche. Compare with *specialist*.

genotype: The genetic constitution of an individual. Compare with *phenotype*.

heritable (*sensu lato*, as employed by early biologists and as used in this book): The ability of information to be passed between individuals or lineages. (Heritable, *sensu stricto*, restricts the meaning to vertical transmission of genetic information.)

hermaphroditism: The state of having both male and female reproductive organs in the same individual. *Simultaneous hermaphrodites* are male and female at the same time; *sequential hermaphrodites* become one sex after being the other.

hypothesis: A falsifiable explanation for an observed pattern. Tests of a hypothesis generate data, to determine if predictions of the hypothesis are evident. Proper science is hypothesis-driven, not data-driven.

intuition: An unconscious conclusion that can inform the conscious mind.

mating system: The pattern of matings between individuals of a population, including particularly the typical number of simultaneous mates members of each sex has.

monogamy: A type of mating system in which one male mates with one female, either for a breeding season or for life. Compare with *polygyny*.

MRCA (most recent common ancestor): That ancestral organism through whom two clades are most closely related.

naturalistic fallacy: An argument that concludes that if something is natural, it is also how things ought to be, effectively applying moral judgment to or inferring it from nature. Closely related to, and for most nonphilosophers interchangeable with, both the is-ought fallacy and the appeal to nature fallacy.

niche: The set of circumstances to which an organism is adapted.

non-zero-sum: An opportunity in which a benefit for one individual does not necessarily come with a cost to conspecifics. Compare with *zero-sum*.

Omega principle (as introduced in this book):

1. Epigenetic phenomena (*sensu lato*) are evolutionarily superior to genetic phenomena in that they are more rapidly adaptable.

2. Epigenetic phenomena (*sensu lato*) are downstream of genetics, so ultimately, genetics are in control.

paradox: The inability to reconcile two observations. Within the universe, all facts must somehow coexist, and therefore the appearance of paradox suggests an incorrect assumption, or other error in understanding. All truths must reconcile.

phenotype: The observable structural and functional properties of an individual. Compare with *genotype*.

plasticity: The capacity of an organism to vary morphologically, physiologically, or behaviorally as a result of environmental variation or fluctuation.

polygyny: A type of mating system in which one male mates with multiple females. Often colloquially called polygamy, but technically, polygamy can refer to asymmetry in number of sexual partners in either direction, so includes both polygyny (one male, many females—common in vertebrates) and polyandry (one female, many males—very rare). Compare with *monogamy*.

proximate: A mechanistic level of explanation, addressing how a given structure or process functions. Compare with *ultimate*.

selection: A process that causes one pattern to become more common than an alternative pattern. Not inherently biotic.

sensu lato: "In the broad sense," from the Latin. Along with *sensu stricto*, originally used to distinguish between names for taxonomic groupings over which there is disagreement in membership. Used in this book to indicate broader, more inclusive meanings of terms more generally.

sensu stricto: "In the narrow sense." See *sensu lato*.

specialist: A species or an individual with narrow tolerances, or adopted to a very narrow niche. Compare with *generalist*.

Sucker's Folly: The tendency of a concentrated short-term benefit not only to obscure risk and long-term cost, but also to drive acceptance even when the net analysis is negative.

theory of mind: The ability to infer mental states—such as beliefs, emotions, or knowledge—of others, especially when they are different from one's own.

trade-off: An obligate, negative relationship between two desirable characteristics. Three types: allocation, design constraint, and statistical.

ultimate: An evolutionary level of explanation, addressing *why* a given structure or process is the way that it is. Compare with *proximate*.

WEIRD: Of **W**estern **E**ducated **I**ndustrialized **R**ich and **D**emocratic countries.

zero-sum: An opportunity in which a benefit for one individual results in an equivalent cost to conspecifics. Compare with *non-zero-sum*.

Recommended Further Reading

Chapter 1: The Human Niche

Dawkins, R., 1976. *The Selfish Gene*. New York: Oxford University Press.

Mann, C. C., 2005. *1491: New Revelations of the Americas before Columbus*. New York: Alfred A. Knopf.

Meltzer, D. J., 2009. *First Peoples in a New World: Colonizing Ice Age America*. Berkeley: University of California Press.

Chapter 2: A Brief History of the Human Lineage

Dawkins, R., and Wong, Y., 2004. *The Ancestor's Tale: A Pilgrimage to the Dawn of Evolution*. New York: Houghton Mifflin.

Shostak, M., 2009. *Nisa: The Life and Words of a !Kung Woman*. Cambridge, MA: Harvard University Press.

Shubin, N., 2008. *Your Inner Fish: A Journey into the 3.5-Billion-Year History of the Human Body*. New York: Vintage.

Chapters 3 and 4: Ancient Bodies, Modern World; Medicine

Burr, C., 2004. *The Emperor of Scent: A True Story of Perfume and Obsession*. New York: Random House.

Lieberman, D., 2014. *The Story of the Human Body: Evolution, Health, and Disease*. New York: Vintage.

Muller, J. Z., 2018. *The Tyranny of Metrics*. Princeton, NJ: Princeton University Press.

Nesse, R. M., and Williams, G. C., 1996. *Why We Get Sick: The New Science of Darwinian Medicine*. New York: Vintage.

Chapter 5: Food

Nabhan, G. P., 2013. *Food, Genes, and Culture: Eating Right for Your Origins*. Washington, D.C.: Island Press.

Pollan, M., 2006. *The Omnivore's Dilemma: A Natural History of Four Meals*. New York: Penguin Press.

Wrangham, R., 2009. *Catching Fire: How Cooking Made Us Human*. New York: Basic Books.

Chapter 6: Sleep

Walker, M., 2017. *Why We Sleep: Unlocking the Power of Sleep and Dreams*. New York: Scribner.

Chapter 7: Sex and Gender

Buss, D. M., 2016. *The Evolution of Desire: Strategies of Human Mating*. New York: Basic Books.

Low, B. S., 2015. *Why Sex Matters: A Darwinian Look at Human Behavior*. Princeton, NJ: Princeton University Press.

Chapter 8: Parenthood and Relationship

Hrdy, S. B., 1999. *Mother Nature: A History of Mothers, Infants, and Natural Selection*. New York: Pantheon.

Junger, S., 2016. *Tribe: On Homecoming and Belonging*. New York: Twelve.

Shenk, J. W., 2014. *Powers of Two: How Relationships Drive Creativity*. New York: Houghton Mifflin Harcourt.

Chapter 9: Childhood

Gray, P., 2013. *Free to Learn: Why Unleashing the Instinct to Play Will Make Our Children Happier, More Self-Reliant, and Better Students for Life*. New York: Basic Books.

Lancy, D. F., 2014. *The Anthropology of Childhood: Cherubs, Chattel, Changelings*. Cambridge: Cambridge University Press.

Chapter 10: School

Crawford, M. B., 2009. *Shop Class as Soulcraft: An Inquiry into the Value of Work*. New York: Penguin Press.

Gatto, J. T., 2010. *Weapons of Mass Instruction: A Schoolteacher's Journey through the Dark World of Compulsory Schooling*. Gabriola Island: New Society Publishers.

Jensen, D., 2005. *Walking on Water: Reading, Writing, and Revolution*. White River Junction, VT: Chelsea Green Publishing.

Chapter 11: Becoming Adults

de Waal, F., 2019. *Mama's Last Hug: Animal Emotions and What They Tell Us about Ourselves*. New York: W. W. Norton.

Kotler, S., and Wheal, J., 2017. *Stealing Fire: How Silicon Valley, the Navy SEALs, and Maverick Scientists Are Revolutionizing the Way We Live and Work*. New York: Harper-Collins.

Lukianoff, G., and Haidt, J., 2019. *The Coddling of the American Mind: How Good Intentions and Bad Ideas Are Setting Up a Generation for Failure*. New York: Penguin Books.

Chapter 12: Culture and Consciousness

Cheney, D. L., and Seyfarth, R. M., 2008. *Baboon Metaphysics: The Evolution of a Social Mind*. Chicago: University of Chicago Press.

Ehrenreich, B., 2007. *Dancing in the Streets: A History of Collective Joy*. New York: Metropolitan Books.

Chapter 13: The Fourth Frontier

Alexander, R. D., 1990. *How Did Humans Evolve? Reflections on the Uniquely Unique Species*. Ann Arbor, MI: Museum of Zoology, University of Michigan, Special Publication No. 1.

Diamond, J. M., 1998. *Guns, Germs and Steel: A Short History of Everybody for the Last 13,000 Years*. New York: Random House.

Sapolsky, R. M., 2017. *Behave: The Biology of Humans at Our Best and Worst*. New York: Penguin Press.

More technical texts that are nonetheless excellent include:

Jablonka, E., and Lamb, M. J., 2014. *Evolution in Four Dimensions: Genetic, Epigenetic, Behavioral, and Symbolic Variation in the History of Life*. Revised edition. Cambridge, MA: MIT Press.

West-Eberhard, M. J., 2003. *Developmental Plasticity and Evolution*. New York: Oxford University Press.

Notes

Introduction

1. See Weinstein, E., 2021. "A Portal Special Presentation—Geometric Unity: A First Look." YouTube video, April 2, 2021. https://youtu.be/Z7rd04KzLcg.
2. There are actually three closely related logical fallacies, the distinctions between which philosophers like to chide the rest of us about when we use them imprecisely: the naturalistic fallacy, the appeal to nature fallacy, and the is-ought fallacy.

Chapter 1: The Human Niche

1. Tamm, E., et al., 2007. Beringian standstill and spread of Native American founders. *PloS One*, 2(9): e829.
2. This is still a somewhat controversial claim, but this nonprimary article lays out some of the evidence well: Wade, L., 2017. On the trail of ancient mariners. *Science*, 357(6351): 542–545.
3. Carrara, P. E., Ager, T. A., and Baichtal, J. F., 2007. Possible refugia in the Alexander Archipelago of southeastern Alaska during the late Wisconsin glaciation. *Canadian Journal of Earth Sciences*, 44(2): 229–244.
4. When the Americas were first peopled is the stuff of legend. Here are just three peer-reviewed articles using different evidence in support of a Beringian arrival in the New World at least sixteen thousand years ago:

     Dillehay, T. D., et al., 2015. New archaeological evidence for an early human presence at Monte Verde, Chile. *PloS One*, 10(11): e0141923; Llamas, B., et al., 2016. Ancient mitochondrial DNA provides high-resolution time scale of the peopling of the Americas. *Science Advances*, 2(4): e1501385; Davis, L. G., et al., 2019. Late Upper Paleolithic occupation at Cooper's Ferry, Idaho, USA, ~16,000 years ago. *Science*, 365(6456): 891–897.
5. Some of the evidence for an even earlier peopling of the Americas comes from cultural artifacts of high-altitude caves in Mexico: Ardelean, C. F., et al., 2020. Evidence of human occupation in Mexico around the Last Glacial Maximum. *Nature*, 584(7819): 87–92; Becerra-Valdivia, L., and Higham, T., 2020. The timing and effect of the earliest human arrivals in North America: *Nature*, 584(7819): 93–97.

6. These early Americans no doubt had fished the cold ocean waters as they made their way down the coast from Beringia, but now many of them made their living on land, developing new skills and technologies. Perhaps they were itinerant, spreading down the coast, then out across the landscape, fan-like, before making permanent settlements. Perhaps they spent some seasons hunkered down, and moved when it was easier to do so—the food more abundant, the climate less dangerous. Fresh water would have been limiting for them, necessary, as it is for all life, and so they would likely have clustered around lakes and streams.

   They would have come upon rivers that filled with salmon every year. The Beringians likely fished for salmon while still in Beringia, and perhaps the technology that they developed there, in the northern flats, kept them in fish on their way down the west coast of North America. Maybe salmon populations returned to rivers that ran out to sea under thin spots in the ice sheets, and it was the salmon that led the Beringians south, the journey being less of a leap of faith: so long as there were fish, there was life. Or perhaps the technology needed to change as they went south, as the geology and rivers shifted with latitude, and some populations forgot their salmon fishing ways, for a while. Perhaps that cultural memory of salmon fishing was latent, just below the surface.

7. At least not on Earth.

8. In *Greenes, Groats-worth of Witte, Bought with a Million of Repentance*, a pamphlet published in the name of deceased playwright Robert Greene. 1592.

9. Humans are extraordinarily extraordinary, and also uniquely unique: Alexander, R. D., 1990. *How Did Humans Evolve? Reflections on the Uniquely Unique Species*. Ann Arbor, MI: Museum of Zoology, University of Michigan, Special Publication No. 1.

10. The funny thing about paradoxes is that in an important sense, they can't be real. There can't be any true contradictions within the structure of the universe—all truths must somehow reconcile. This is the assumption that undergirds the scientific endeavor itself. Science is the search for the insights that reconcile paradoxes. As Niels Bohr once said, "How wonderful that we have met a paradox. Now we have some hope of making progress."

11. See, for instance, any number of works on flow by Mihály Csíkzentmihályi.

12. The Sucker's Folly is related to the economic concept of discounting, as well as to that of the "progress trap," which is well laid out in O'Leary, D. B., 2007. *Escaping the Progress Trap*. Montreal: Geozone Communications.

13. Most Recent Common Ancestor (MRCA) is a term of art in phylogenetic systematics, but to reduce jargon, we are using lowercase, as the phrase means what it seems to.

14. In part, people object when evolutionary theory is invoked to explain behavior or culture because soon after its discovery, evolutionary theory was misappropriated, and used to justify regressive social conclusions and policies under the umbrella of pseudoscientific "social Darwinism." The word *lineage* has, similarly, been used to unsavory ends. Such errors in thinking produced, for instance, the belief by rich Gilded Age Americans that their wealth was an indicator of their evolutionary superiority; more than a century of forced sterilizations throughout America; and

Nazism. Such mistakes reveal the naturalistic fallacy in action: having understood correctly that we are the products of evolution, it is an easy, if wrong, step for those currently in power to claim that their current power is proof of their superiority (error one), and not just now but forever (error two). For more on this, see: N. K. Nittle, 2021. The government's role in sterilizing women of color. ThoughtCo. https://www.thoughtco.com/u-s-governments-role-sterilizing-women-of-color-2834600; *Radiolab*—"G: Unfit" podcast episode, first aired July 17, 2019, download and transcript available at https://www.wnycstudios.org/podcasts/radiolab/articles/g-unfit.

15. The distinction between individual and population is critical. Being a member of some *population*—women, Europeans, right-handers—specifies very few hard-and-fast truths about the *individual*, while making many other characteristics more or less likely for individual members of that group.

16. Dawkins, R. 1976. *The Selfish Gene* (30th anniversary ed. [2006]). New York: Oxford University Press, 192.

17. We first introduced the concept of the Omega principle outside of the classroom at a Baumann Foundation event on "Being Human" in San Francisco in July 2014, at the invitation of Peter Baumann. Our presentation was nine hours long over two days, and included many of the ideas herein. We presented similar work to The Leakey Foundation in April 2015. We remain grateful to both for the opportunities.

Chapter 2: A Brief History of the Human Lineage

1. (Nearly) all of these examples are from Brown, D., 1991. "The Universal People." In *Human Universals*. New York: McGraw Hill.

2. Brunet, T., and King, N., 2017. The origin of animal multicellularity and cell differentiation. *Developmental Cell*, 43(2): 124–140.

3. The Paleognaths (literally: "old jaws") include most of the flightless clades of birds, but molecular evidence suggests that there were several evolutions of flightlessness, not that they all evolved from a flightless ancestor. Mitchell, K. J., et al., 2014. Ancient DNA reveals elephant birds and kiwi are sister taxa and clarifies ratite bird evolution. *Science*, 344(6186): 898–900.

4. Espinasa, L., Rivas-Manzano, P., and Pérez, H. E., 2001. A new blind cave fish population of genus *Astyanax*: Geography, morphology and behavior. *Environmental Biology of Fishes*, 62(1–3): 339–344.

5. Welch, D. B. M., and Meselson, M., 2000. Evidence for the evolution of bdelloid rotifers without sexual reproduction or genetic exchange. *Science*, 288(5469): 1211–1215.

6. Gladyshev, E., and Meselson, M., 2008. Extreme resistance of bdelloid rotifers to ionizing radiation. *Proceedings of the National Academy of Sciences*, 105(13): 5139–5144.

7. In fact, our lineage has probably been sexually reproducing for far longer than that—between one and two billion years by many estimates. Five hundred million years is a conservative estimate that is roughly equivalent to when vertebrates first evolved.

8. Dunn, C. W., et al., 2014. Animal phylogeny and its evolutionary implications. *Annual Review of Ecology, Evolution, and Systematics*, 45: 371–395.

9. Dunn et al., 2014.

10. Zhu, M., et al., 2013. A Silurian placoderm with osteichthyan-like marginal jaw bones. *Nature*, 502(7470): 188–193.

11. For more on this kind of thinking, see Weinstein, B., 2016. On being a fish. *Inference: International Review of Science*, 2(3): September 2016. https://inference-review.com/article/on-being-a-fish.

12. Springer, M. S., et al., 2003. Placental mammal diversification and the Cretaceous–Tertiary boundary. *Proceedings of the National Academy of Sciences*, 100(3): 1056–1061; Foley, N. M., Springer, M. S., and Teeling, E. C., 2016. Mammal madness: Is the mammal tree of life not yet resolved? *Philosophical Transactions of the Royal Society B: Biological Sciences*, 371(1699): 1056–1061.

13. *Character* is a term of art in systematics, the science of discovering the deep history of organismal relationships. In common parlance, *characteristic* is a fairly close but imperfect approximation.

14. Known as Carrier's constraint.

15. Some of these early mammal adaptations include the four-chambered heart (circulatory); diaphragm (respiratory); parasagittal gait (locomotory); unique anatomy of our inner ear (auditory), which is related to having a single bone in the lower jaw, which in combination with temporal fenestration as attachment points for jaw muscles allows for stronger bite force; and the loop of Henle in the kidney, which refines our excretion of nitrogenous waste.

16. Renne, P. R., et al., 2015. State shift in Deccan volcanism at the Cretaceous-Paleogene boundary, possibly induced by impact. *Science*, 350(6256): 76–78.

17. For example, Silcox, M. T., and López-Torres, S., 2017. Major questions in the study of primate origins. *Annual Review of Earth and Planetary Sciences*, 45: 113–137.

18. Bret is not so sure about this claim.

19. See, for example, Steiper, M. E., and Young, N. M., 2006. Primate molecular divergence dates. *Molecular Phylogenetics and Evolution*, 41(2): 384–394; Stevens, N. J., et al., 2013. Palaeontological evidence for an Oligocene divergence between Old World monkeys and apes. *Nature*, 497(7451): 611.

20. See, for example, Wilkinson, R. D., et al., 2010. Dating primate divergences through an integrated analysis of palaeontological and molecular data. *Systematic Biology*, 60(1): 16–31.

21. Hobbes, T., 1651. *Leviathan*. Chapter XIII: "Of the Natural Condition of Mankind as Concerning Their Felicity and Misery."

22. Niemitz, C., 2010. The evolution of the upright posture and gait—a review and a new synthesis. *Naturwissenschaften*, 97(3): 241–263.

23. Preuschoft, H., 2004. Mechanisms for the acquisition of habitual bipedality: Are there biomechanical reasons for the acquisition of upright bipedal posture? *Journal of Anatomy*, 204(5): 363–384.

24. Hewes, G. W., 1961. Food transport and the origin of hominid bipedalism. *American Anthropologist*, 63(4): 687–710.

25. See, for example, Provine, R. R., 2017. Laughter as an approach to vocal evolution: The bipedal theory. *Psychonomic Bulletin & Review*, 24(1): 238–244.

26. Alexander, R. D., 1990. *How Did Humans Evolve? Reflections on the Uniquely Unique Species*. Ann Arbor, MI: Museum of Zoology, University of Michigan. Special Publication No. 1.

27. See, for example, Conard, N. J., 2005. "An Overview of the Patterns of Behavioural Change in Africa and Eurasia during the Middle and Late Pleistocene." In *From Tools to Symbols: From Early Hominids to Modern Humans*, d'Errico, F., Backwell, L., and Malauzat, B., eds. New York: NYU Press, 294–332.

28. Aubert, M., et al., 2014. Pleistocene cave art from Sulawesi, Indonesia. *Nature*, 514 (7521): 223.

29. Hoffmann, D. L., et al., 2018. U-Th dating of carbonate crusts reveals Neandertal origin of Iberian cave art. *Science*, 359(6378): 912–915.

30. Lynch, T. F., 1989. Chobshi cave in retrospect. *Andean Past*, 2(1): 4.

31. Stephens, L., et al., 2019. Archaeological assessment reveals Earth's early transformation through land use. *Science*, 365(6456): 897–902.

32. Using birth and death records of people famous enough during their lifetimes to have their births and deaths recorded, scientists recently mapped cultural centers since the time of the Roman Empire. Schich, M., et al., 2014. A network framework of cultural history. *Science*, 345(6196): 558–562.

Chapter 3: Ancient Bodies, Modern World

1. Segall, M., Campbell, D., and Herskovits, M. J., 1966. *The Influence of Culture on Visual Perception*. New York: Bobbs-Merrill.

2. Hubel, D. H., and Wiesel, T. N., 1964. Effects of monocular deprivation in kittens. *Naunyn-Schmiedebergs Archiv for Experimentelle Pathologie und Pharmakologie*, 248: 492–497.

3. See, for instance, Henrich, J., Heine, S. J., and Norenzayan, A., 2010. The weirdest people in the world? *Behavioral and Brain Sciences*, 33(2–3): 61–83; Gurven, M. D., and Lieberman, D. E., 2020. WEIRD bodies: Mismatch, medicine and missing diversity. *Evolution and Human Behavior*, 41(2020): 330–340.

4. Holden, C., and Mace, R., 1997. Phylogenetic analysis of the evolution of lactose digestion in adults. *Human Biology*, 81(5/6): 597–620.

5. Flatz, G., 1987. "Genetics of Lactose Digestion in Humans." In *Advances in Human Genetics*. Boston: Springer, 1–77.

6. Segall, Campbell, and Herskovits, *Influence of Culture*, 32.

7. Owen, N., Bauman, A., and Brown, W., 2009. Too much sitting: A novel and important predictor of chronic disease risk? *British Journal of Sports Medicine*, 43(2): 81–83.

8. Metchnikoff, E., 1903. *The Nature of Man,* as cited in Keith, A., 1912. The functional nature of the caecum and appendix. *British Medical Journal,* 2: 1599–1602.

9. Keith, Functional nature of the caecum and appendix.

10. In the case of polar bears, the advantage of being white surely drove the loss of pigment in their fur. In the case of naked mole rats, the absence of hair may provide an advantage, such as resistance to parasites, or may have been driven by savings alone, as naked mole rats live their lives in insulated, underground environments.

11. Berry, R. J. A., 1900. The true caecal apex, or the vermiform appendix: Its minute and comparative anatomy. *Journal of Anatomy and Physiology,* 35(Part 1): 83–105.

12. Laurin, M., Everett, M. L., and Parker, W., 2011. The cecal appendix: One more immune component with a function disturbed by post-industrial culture. *Anatomical Record: Advances in Integrative Anatomy and Evolutionary Biology,* 294(4): 567–579.

13. Bollinger, R. R., et al., 2007. Biofilms in the large bowel suggest an apparent function of the human vermiform appendix. *Journal of Theoretical Biology,* 249(4): 826–831.

14. Boschi-Pinto, C., Velebit, L., and Shibuya, K., 2008. Estimating child mortality due to diarrhoea in developing countries. *Bulletin of the World Health Organization,* 86: 710–717.

15. Laurin, Everett, and Parker, The cecal appendix, 569.

16. Bickler, S. W., and DeMaio, A., 2008. Western diseases: Current concepts and implications for pediatric surgery research and practice. *Pediatric Surgery International,* 24(3): 251–255.

17. Rook, G. A., 2009. Review series on helminths, immune modulation and the hygiene hypothesis: The broader implications of the hygiene hypothesis. *Immunology,* 126(1): 3–11.

18. Chesterton, G. K., 1929. "The Drift from Domesticity." In *The Thing.* Aeterna Press.

19. The adaptive landscapes metaphor is often described as ranges of mountain peaks and valleys, but it is easier to understand their evolutionary implications if we consider a transparent sheet of ice on the surface of a pond. Air bubbles floating through the water get trapped under the ice, driven up by gravity, finding the high spots.

   Those peaks represent ecological opportunities, and the bubbles represent creatures evolving to exploit these opportunities through evolutionary adaptation. The force of gravity represents the force of selection that refines organisms, fitting them to their niche. The bigger the peak, the greater the ecological opportunity it represents. The thick-ice "valleys" represent obstacles that block bubbles from moving between peaks.

   This metaphor is useful in conceptualizing evolutionary dynamics, especially where the process is counterintuitive. Consider, for example, the case of a small bubble trapped in a low peak that is right next to a taller peak. Taller peaks represent better opportunities, so you'd expect selection to move things from low peaks to the higher ones. But that expectation is wrong. Selection has no means to make creatures worse in order to improve them down the road, just as gravity cannot move a bubble deeper into the water in order to have it rise higher, elsewhere. Some other force must be responsible for all downhill motion—like someone jumping on the ice. Further, the likelihood that a bubble will move from a low peak to a high one is not related to the difference in their heights, as you might expect. Rather, it is related to the depth of the valley that separates them. The deeper the valley, the greater the barrier to the discovery of opportunity.

The metaphor was first introduced here: Wright, S. 1932. The roles of mutation, inbreeding, crossbreeding, and selection in evolution. *Proceedings of the Sixth International Congress of Genetics*, 1: 356–366.

20. There is a third type, the statistical trade-off, but it is not truly a trade-off. Rather, it is an observation that individuals with multiple, uncommon characteristics are rarer than individuals with a single uncommon characteristic. You want a gray dog? Fine. You want a giant dog? Sure thing. You want a giant, gray dog? Much tougher to get than either a giant or a gray dog.

For further expansion of the adaptive landscapes metaphor as it applies to trade-offs, and including a reimagining of the landscape as a volume that fills as individuals find opportunities, which allows it to explain both diversification of forms and exploration of new space, literal and metaphorical, see Weinstein, B. S., 2009. "Evolutionary Trade-offs: Emergent Constraints and Their Adaptive Consequences." A dissertation submitted in partial fulfillment of the requirements for the degree of Doctor of Philosophy (Biology), University of Michigan.

21. Here we are referring to fishy fish (e.g., salmon, angelfish, gobies) to distinguish them from all fish, a clade to which we belong. See chapter 2, and also Weinstein, B., On being a fish. *Inference: International Review of Science*, 2(3): September 2016.

22. Schrank, A. J., Webb, P. W., and Mayberry, S., 1999. How do body and paired-fin positions affect the ability of three teleost fishes to maneuver around bends? *Canadian Journal of Zoology*, 77(2): 203–210.

23. The point being that even two qualities that do not appear to be connected are in trade-off relationship with each other. See Weinstein, "Evolutionary Trade-offs."

24. A term coined here: Dawkins, R., 1982. *The Extended Phenotype*. Oxford: Oxford University Press.

25. Another type of photosynthesis, C4, separates in space what CAM separates in time, and is, like CAM, both an adaptation to hot, dry conditions and more metabolically expensive than C3 photosynthesis.

26. This was told to Bret by our professor George Estabrook many years ago.

27. For a fantastic scientific story, see Burr, C., 2004. *The Emperor of Scent: A True Story of Perfume and Obsession*. New York: Random House.

28. Feinstein, J. S., et al., 2013. Fear and panic in humans with bilateral amygdala damage. *Nature Neuroscience*, 16(3): 270–272.

Chapter 4: Medicine

1. For a relatively early take, which has become a classic, see Nesse, R., and Williams, G., 1996. *Why We Get Sick: The New Science of Darwinian Medicine.* New York: Vintage.

2. Tenger-Trolander, A., et al., 2019. Contemporary loss of migration in monarch butterflies. *Proceedings of the National Academy of Sciences,* 116(29): 14671–14676.

3. Britt, A., et al., 2002. Diet and feeding behaviour of *Indri indri* in a low-altitude rain forest. *Folia Primatologica,* 73(5): 225–239.

4. The first of Hayek's essays on the topic is: Hayek, F. V., 1942. Scientism and the study of society. Part I. *Economica,* 9(35): 267–291. See also Hayek, F. A., 1945. The use of knowledge in society. *The American Economic Review,* 35(4): 519–530.

5. Aviv, R., 2019. Bitter pill. *New Yorker,* April 8, 2019. https://www.newyorker.com/magazine/2019/04/08/the-challenge-of-going-off-psychiatric-drugs.

6. See, for example, Choi, K. W., et al., 2020. Physical activity offsets genetic risk for incident depression assessed via electronic health records in a biobank cohort study. *Depression and Anxiety,* 37(2): 106–114.

7. Tomasi, D., Gates, S., and Reyns, E., 2019. Positive patient response to a structured exercise program delivered in inpatient psychiatry. *Global Advances in Health and Medicine,* 8: 1–10.

8. Gritters, J., "Is CBG the new CBD?," *Elemental,* on Medium. July 8, 2019. https://elemental.medium.com/is-cbg-the-new-cbd-6de59e568008.

9. Mann, C., 2020. Is there still a good case for water fluoridation?, *Atlantic,* April 2020. https://www.theatlantic.com/magazine/archive/2020/04/why-fluoride-water/606784.

10. Choi, A. L., et al., 2015. Association of lifetime exposure to fluoride and cognitive functions in Chinese children: A pilot study. *Neurotoxicology and Teratology,* 47: 96–101.

11. Malin, A. J., et al., 2018. Fluoride exposure and thyroid function among adults living in Canada: Effect modification by iodine status. *Environment International,* 121: 667–674.

12. Damkaer, D. M., and Dey, D. B., 1989. Evidence for fluoride effects on salmon passage at John Day Dam, Columbia River, 1982–1986. *North American Journal of Fisheries Management,* 9(2): 154–162.

13. Abdelli, L. S., Samsam, A., and Naser, S. A., 2019. Propionic acid induces gliosis and neuro-inflammation through modulation of PTEN/AKT pathway in autism spectrum disorder. *Scientific Reports,* 9(1): 1–12.

14. Autier, P., et al., 2014. Vitamin D status and ill health: A systematic review. *Lancet Diabetes & Endocrinology,* 2(1): 76–89.

15. Jacobsen, R., 2019. Is sunscreen the new margarine? *Outside Magazine,* January 10, 2019, https://www.outsideonline.com/2380751/sunscreen-sun-exposure-skin-cancer-science.

16. Lindqvist, P. G., et al., 2016. Avoidance of sun exposure as a risk factor for major causes of death: A competing risk analysis of the melanoma in southern Sweden cohort. *Journal of Internal Medicine*, 280(4): 375–387.

17. Marchant, J., 2018. When antibiotics turn toxic. *Nature*, 555(7697): 431–433.

18. Mayr, E., 1961. Cause and effect in biology. *Science*, 134(3489): 1501–1506.

19. Dobzhansky, D., 1973. Nothing in Biology Makes Sense except in the Light of Evolution. *The American Biology Teacher*, 35(3): 125–129.

20. In response to this confusing political rhetoric, we began livestreaming in late March 2020, largely on the topic of COVID-19 for the first two months. *The Evolutionary Lens*, the cohosted (by us) branch of Bret's *DarkHorse* podcast, demonstrates evolutionary thinking about this and other contemporary topics every week.

21. Among many other reasons, evidence is growing that exercise mitigates some mood disorders. See, for example, Choi, K. W., et al., 2020. Physical activity offsets genetic risk for incident depression assessed via electronic health records in a biobank cohort study. *Depression and Anxiety*, 37(2): 106–114.

22. Holowka, N. B., et al., 2019. Foot callus thickness does not trade off protection for tactile sensitivity during walking. *Nature*, 571(7764): 261–264.

23. Jacka, F. N., et al., 2017. A randomised controlled trial of dietary improvement for adults with major depression (the ("SMILES" trial). *BMC Medicine*, 15(1): 23.

24. Lieberman, D., 2014. *The Story of the Human Body: Evolution, Health, and Disease*. New York: Vintage.

Chapter 5: Food

1. Wrangham, R., 2009. *Catching Fire: How Cooking Made Us Human*. New York: Basic Books, 80.

2. Craig, W. J., 2009. Health effects of vegan diets. *American Journal of Clinical Nutrition*, 89(5): 1627S–1633S.

3. Wadley, L., et al., 2020. Cooked starchy rhizomes in Africa 170 thousand years ago. *Science*, 367(6473): 87–91.

4. Field, H., 1932. Ancient wheat and barley from Kish, Mesopotamia. *American Anthropologist*, 34(2): 303–309.

5. Kaniewski, D., et al., 2012. Primary domestication and early uses of the emblematic olive tree: Palaeobotanical, historical and molecular evidence from the Middle East. *Biological Reviews*, 87(4): 885–899.

6. Bellwood, P. S., 2005. *First Farmers: The Origins of Agricultural Societies*. Oxford: Blackwell Publishing, 97.

7. Struhsaker, T. T., and Hunkeler, P., 1971. Evidence of tool-using by chimpanzees in the Ivory Coast. *Folia Primatologica*, 15(3–4): 212–219.

8. Goodall, J., 1964. Tool-using and aimed throwing in a community of free-living chimpanzees. *Nature*, 201(4926): 1264–1266.

9. Marlowe, F. W., et al., 2014. Honey, Hadza, hunter-gatherers, and human evolution. *Journal of Human Evolution*, 71: 119–128.

10. Harmand, S., et al., 2015. 3.3-million-year-old stone tools from Lomekwi 3, west Turkana, Kenya. *Nature*, 521(7552): 310–326.

11. De Heinzelin, J., et al., 1999. Environment and behavior of 2.5-million-year-old Bouri hominids. *Science*, 284(5414): 625–629.

12. Bellomo, R. V., 1994. Methods of determining early hominid behavioral activities associated with the controlled use of fire at FxJj 20 Main, Koobi Fora, Kenya. *Journal of Human Evolution*, 27(1-3): 173–195. Also see Wrangham, R. W., et al., 1999. The raw and the stolen: Cooking and the ecology of human origins. *Current Anthropology*, 40(5): 567–594.

13. Tylor, E. B., 1870. *Researches into the Early History of Mankind and the Development of Civilization*. London: John Murray, 231–239.

14. Darwin, C., 1871. *The Descent of Man, and Selection in Relation to Sex*. London: Murray, 415.

15. Wrangham, *Catching Fire*.

16. In 1860, European explorers in Australia were on the brink of starvation when they asked the local Yandruwandha aborigines for help. The local people pointed the explorers to the roots of the abundant nardoo plant, which the locals pounded into flour, washed, and cooked. Two of the Europeans omitted the washing and cooking, became weak, and died. One of their companions, who ate with and as the Yandruwandha did, was in excellent condition when rescued ten weeks later (as told in Wrangham, *Catching Fire*, 35).

17. Wrangham, *Catching Fire*, 138–142.

18. Tylor, Researches into the Early History of Mankind, 233.

19. Tylor, Researches into the Early History of Mankind, 263.

20. This evolutionary shorthand ("seeds don't want to be eaten") will strike some as odd, as if we're attributing consciousness or will to seeds. This is not the intention. A wordier version of the same sentiment would read: "Seeds are not produced by plants with the intention that they be eaten."

21. Toniello, G., et al., 2019. 11,500 y of human–clam relationships provide long-term context for intertidal management in the Salish Sea, British Columbia. *Proceedings of the National Academy of Sciences*, 116(44): 22106–22114.

22. Bellwood, *First Farmers*.

23. Arranz-Otaegui, A., et al., 2018. Archaeobotanical evidence reveals the origins of bread 14,400 years ago in northeastern Jordan. *Proceedings of the National Academy of Sciences*, 115(31): 7295–7930.

24. Brown, D., 1991. *Human Universals*. New York: McGraw Hill.

25. Wu, X., et al., 2012. Early pottery at 20,000 years ago in Xianrendong Cave, China. *Science* 336(6089): 1696–1700.

26. Braun, D. R., et al., 2010. Early hominin diet included diverse terrestrial and aquatic animals 1.95 Ma in East Turkana, Kenya. *Proceedings of the National Academy of Sciences*, 107(22): 10002–10007.

27. Archer, W., et al., 2014. Early Pleistocene aquatic resource use in the Turkana Basin. *Journal of Human Evolution*, 77(2014): 74–87.

28. Marean, C. W., et al., 2007. Early human use of marine resources and pigment in South Africa during the Middle Pleistocene. *Nature,* 449(7164): 905–908.

29. Koops, K., et al., 2019. Crab-fishing by chimpanzees in the Nimba Mountains, Guinea. *Journal of Human Evolution,* 133: 230–241.

30. Pollan, M., 2006. *The Omnivore's Dilemma: A Natural History of Four Meals.* New York: Penguin Press.

31. As Michael Pollan says in *The Omnivore's Dilemma,* if your grandmother wouldn't recognize it as food, it's not food. For pregnant women, though, it's not just as simple as eating real food, as fetuses are also susceptible to pathogens from food that is real and that healthy adults can normally eat. So, alas, pregnancy is not the moment to eat goat or sheep cheese or any mold-ripened or raw cheese, or salami or most deli meats.

32. Collection of wild honey is a highly male-gendered activity across cultures, as reported in Murdock, G. P., and Provost, C., 1973. Factors in the division of labor by sex: A cross-cultural analysis. *Ethnology,* 12(2): 203–225, as well as in Marlowe et al., Honey, Hadza, hunter-gatherers.

Chapter 6: Sleep

1. Walker, M., 2017. *Why We Sleep: Unlocking the Power of Sleep and Dreams.* New York: Scribner, 56–57.

2. Tidally locked planets, which are half in permanent day and half in permanent night because half the planet always faces its sun, are unlikely to support life. The differences between the two halves is so extreme that there is unlikely to be a goldilocks zone for planets that are tidally locked.

3. Walker, *Why We Sleep,* 46–49. Different researchers categorize sleep differently. In his book, Walker uses "REM" and "NREM" (non-REM), but also clarifies that NREM's four stages are further divided: NREM stages 3 and 4 are "slow-wave sleep"; NREM stages 1 and 2 are shallow and light in comparison.

4. Shein-Idelson, M., et al., 2016. Slow waves, sharp waves, ripples, and REM in sleeping dragons. *Science,* 352(6285): 590–595.

5. Martin-Ordas, G., and Call, J., 2011. Memory processing in great apes: The effect of time and sleep. *Biology Letters,* 7(6): 829–832.

6. Walker, *Why We Sleep,* 133.

7. Wright, G. A., et al., 2013. Caffeine in floral nectar enhances a pollinator's memory of reward. *Science,* 339(6124): 1202–1204.

8. Phillips, A. J. K., et al., 2019. High sensitivity and interindividual variability in the response of the human circadian system to evening light. *Proceedings of the National Academy of Sciences,* 116(24): 12019–12024.

9. See, for example, Stevens, R. G., et al., 2013. Adverse health effects of nighttime lighting: Comments on American Medical Association policy statement. *American Journal of Preventive Medicine,* 45(3): 343–346.

10. Hsiao, H. S., 1973. Flight paths of night-flying moths to light. *Journal of Insect Physiology,* 19(10): 1971–1976.

11. Le Tallec, T., Perret, M., and Théry, M., 2013. Light pollution modifies the expression of daily rhythms and behavior patterns in a nocturnal primate. *PloS One*, 8(11): e79250.

12. Gaston, K. J., et al., 2013. The ecological impacts of nighttime light pollution: A mechanistic appraisal. *Biological Reviews*, 88(4): 912–927.

13. Navara, K. J., and Nelson, R. J., 2007. The dark side of light at night: Physiological, epidemiological, and ecological consequences. *Journal of Pineal Research*, 43(3): 215–224.

14. Olini, N., Kurth, S., and Huber, R., 2013. The effects of caffeine on sleep and maturational markers in the rat. *PloS One*, 8(9): e72539.

15. See this remarkable overview of what was already known in 1975 about the limitations of artificial light in keeping people healthy: Wurtman, R. J., 1975. The effects of light on the human body. *Scientific American*, 233(1): 68–79.

16. Park, Y. M. M., et al., 2019. Association of exposure to artificial light at night while sleeping with risk of obesity in women. *JAMA Internal Medicine*, 179(8): 1061–1071.

17. Kernbach, M. E., et al., 2018. Dim light at night: Physiological effects and ecological consequences for infectious disease. *Integrative and Comparative Biology*, 58(5): 995–1007.

Chapter 7: Sex and Gender

1. Association of American Medical Colleges, 2019. *2019 Physician Specialty Data Report: Active Physicians by Sex and Specialty*. Washington, D.C.: AAMC. https://www.aamc.org/data-reports/workforce/interactive-data/active-physicians-sex-and-specialty-2019.

2. Bureau of Labor Statistics, US Department of Labor. Labor Force Statistics from the Current Population Survey. 18. Employed persons by detailed industry, sex, race, and Hispanic or Latino ethnicity. Accessed October 2020, https://www.bls.gov/cps/cpsaat18.htm.

3. Bureau of Labor Statistics. Labor Force Statistics.

4. Eme, L., et al., 2014. On the age of eukaryotes: Evaluating evidence from fossils and molecular clocks. *Cold Spring Harbor Perspectives in Biology*, 6(8): a016139.

5. This is, of course, a bit of an oversimplification. Asexually reproducing organisms can actually do well for themselves even absent a static environment. They deal with stochasticity through mutation and higher reproductive rates. Sexual organisms rely on recombination of tried-and-true genes to maintain an adaptive rate of change relative to their environment. Mutation is still the source of novelty (ultimately) but the cost of mutation is spread out over an entire population, with the good ones getting spread around rather than being limited to each individual lineage. It's all about maintaining an adaptive rate of change relative to the environment: If you are simple, cloning and mutation work. If you are complex, sex is a better bet. Both accomplish the same thing, which is allowing for enough change to match the historic stability of the environment.

6. Notable exceptions are the monotremes, the ~ five species at the base of the mammalian tree that include echidnas and the duck-billed platypus, which have nine or ten sex chromosomes (!). See, for example, Zhou, Y., et al., 2021. Platypus and echidna genomes reveal mammalian biology and evolution. *Nature*, 2021: 1-7.

7. Birds also have GSD (genetic sex determination), but their system evolved independently and is reversed from the mammalian paradigm: males are ZZ (homogametic), females are ZW (heterogametic).

8. As reviewed in Arnold, A. P., 2017. "Sex Differences in the Age of Genetics." In *Hormones, Brain and Behavior*, 3rd ed., Pfaff, D. W., and Joels, M., eds. Cambridge, UK: Academic Press, 33-48.

9. Ferretti, M. T., et al., 2018. Sex differences in Alzheimer disease—the gateway to precision medicine. *Nature Reviews Neurology*, 14: 457-469.

10. Vetvik, K. G., and MacGregor, E. A., 2017. Sex differences in the epidemiology, clinical features, and pathophysiology of migraine. *Lancet Neurology*, 16(1): 76-87.

11. Lynch, W. J., Roth, M. E., and Carroll, M. E., 2002. Biological basis of sex differences in drug abuse: Preclinical and clinical studies. *Psychopharmacology*, 164(2): 121-137.

12. Szewczyk-Krolikowski, K., et al., 2014. The influence of age and gender on motor and non-motor features of early Parkinson's disease: Initial findings from the Oxford Parkinson Disease Center (OPDC) discovery cohort. *Parkinsonism & Related Disorders*, 20(1): 99-105.

13. See, for example: Allen, J. S., et al., 2003. Sexual dimorphism and asymmetries in the gray-white composition of the human cerebrum. *Neuroimage*, 18(4): 880-894; Ingalhalikar, M., et al., 2014. Sex differences in the structural connectome of the human brain. *Proceedings of the National Academy of Sciences*, 111(2): 823-828.

14. Kaiser, T., 2019. Nature and evoked culture: Sex differences in personality are uniquely correlated with ecological stress. *Personality and Individual Differences*, 148: 67-72.

15. Chapman, B. P., et al., 2007. Gender differences in Five Factor Model personality traits in an elderly cohort. *Personality and Individual Differences*, 43(6): 1594-1603.

16. Arnett, A. B., et al., 2015. Sex differences in ADHD symptom severity. *Journal of Child Psychology and Psychiatry*, 56(6): 632-639.

17. See, for example, Altemus, M., Sarvaiya, N., and Epperson, C. N., 2014. Sex differences in anxiety and depression clinical perspectives. *Frontiers in Neuroendocrinology*, 35(3): 320-330; McLean, C. P., et al., 2011. Gender differences in anxiety disorders: Prevalence, course of illness, comorbidity and burden of illness. *Journal of Psychiatric Research*, 45(8): 1027-1035.

18. Su, R., Rounds, J., and Armstrong, P. I., 2009. Men and things, women and people: A meta-analysis of sex differences in interests. *Psychological Bulletin*, 135(6): 859-884.

19. Brown, D., 1991. *Human Universals*. New York: McGraw Hill, 133.

20. Reviewed in Neaves, W. B., and Baumann, P., 2011. Unisexual reproduction among vertebrates. *Trends in Genetics*, 27(3): 81-88.

21. Watts, P. C., et al., 2006. Parthenogenesis in Komodo dragons. *Nature*, 444(7122): 1021-1022.

22. Flame wrasses are reef fish endemic to Hawaii, not, as one of our readers imagined and perhaps hoped for, bipeds from Middle Earth who wear robes and shoot fire. More's the pity.

23. Sullivan, B. K., et al., 1996. Natural hermaphroditic toad (*Bufo microscaphus × Bufo woodhousii*). *Copeia*, 1996(2): 470–472.

24. Grafe, T. U., and Linsenmair, K. E., 1989. Protogynous sex change in the reed frog *Hyperolius viridiflavus*. *Copeia*, 1989(4): 1024–1029.

25. Endler, J. A., Endler, L. C., and Doerr, N. R., 2010. Great bowerbirds create theaters with forced perspective when seen by their audience. *Current Biology*, 20(18): 1679–1684.

26. Alexander, R. D., and Borgia, G., 1979. "On the Origin and Basis of the Male-Female Phenomenon." In *Sexual Selection and Reproductive Competition in Insects*, Blum, M. S., and Blum, N. A., eds. New York: Academic Press. 417–440.

27. Jenni, D. A., and Betts, B. J., 1978. Sex differences in nest construction, incubation, and parental behaviour in the polyandrous American jacana (*Jacana spinosa*). *Animal Behaviour*, 1978(26): 207–218.

28. Claus, R., Hoppen, H. O., and Karg, H., 1981. The secret of truffles: A steroidal pheromone? *Experientia*, 37(11): 1178–1179.

29. Low, B. S., 1979. "Sexual Selection and Human Ornamentation." In *Evolutionary Biology and Human Social Behavior*, Chagnon, N., and Irons, W., eds. Belmont, CA: Duxbury Press, 462–487.

30. Lancaster, J. B., and Lancaster, C. S., 1983. "Parental investment: The hominid adaptation." In *How Humans Adapt: A Biocultural Odyssey*, Ortner, D. J., ed. Washington, D.C.: Smithsonian Institution Press, 33–56.

31. See, for example, Buikstra, J. E., Konigsberg, L. W., and Bullington, J., 1986. Fertility and the development of agriculture in the prehistoric Midwest. *American Antiquity*, 51(3): 528–546.

32. Su, Rounds, and Armstrong, Men and things.

33. Su, Rounds, and Armstrong, Men and things.

34. Reilly, D., 2012. Gender, culture, and sex-typed cognitive abilities. *PloS One*, 7(7): e39904.

35. Deary, I. J., et al., 2003. Population sex differences in IQ at age 11: The Scottish mental survey 1932. *Intelligence*, 31: 533–542.

36. Herrera, A. Y., Wang, J., and Mather, M., 2019. The gist and details of sex differences in cognition and the brain: How parallels in sex differences across domains are shaped by the locus coeruleus and catecholamine systems. *Progress in Neurobiology*, 176: 120–133.

37. Connellan, J., et al., 2000. Sex differences in human neonatal social perception. *Infant Behavior and Development*, 23(1): 113–118.

38. Lancy, D. F., 2014. *The Anthropology of Childhood: Cherubs, Chattel, Changelings*. Cambridge: Cambridge University Press, 258–259.

39. Murdock, G. P., and Provost, C., 1973. Factors in the division of labor by sex: A cross-cultural analysis. *Ethnology*, 12(2): 203–225.

40. Kantner, J., et al., 2019. Reconstructing sexual divisions of labor from fingerprints on Ancestral Puebloan pottery. *Proceedings of the National Academy of Sciences,* 116(25): 12220–12225.

41. Buss, D. M., 1989. Sex differences in human mate preferences: Evolutionary hypotheses tested in 37 cultures. *Behavioral and Brain Sciences,* 12(1): 1–14.

42. Schneider, D. M., and Gough, K., eds., 1961. *Matrilineal Kinship.* Oakland: University of California Press. In particular: Gough, K., "Nayar: Central Kerala," 298–384; Schneider, D. M., "Introduction: The Distinctive Features of Matrilineal Descent Groups," 1–29.

43. See, for example, Trivers, R., 1972. "Parental Investment and Sexual Selection." In *Sexual Selection and the Descent of Man,* Campbell, B., ed. New York: Aldine DeGruyter, 136–179.

44. Buss, D. M., Sex differences in human mate preferences.

45. Buss, D. M., et al., 1992. Sex differences in jealousy: Evolution, physiology, and psychology. *Psychological Science,* 3(4): 251–256.

46. Brickman, J. R., 1978. "Erotica: Sex Differences in Stimulus Preferences and Fantasy Content." A dissertation submitted in partial fulfillment of the requirements for the degree of Doctor of Philosophy, Department of Psychology, University of Manitoba.

47. Three articles that connect the rise of porn with sexual violence against women in otherwise consensual interactions include Julian, K., 2018. The sex recession. *Atlantic,* December 2018. https://www.theatlantic.com/magazine/archive/2018/12/the-sex-recession/573949; Bonnar, M. "I thought he was going to tear chunks out of my skin." BBC News, March 23, 2020. https://www.bbc.com/news/uk-scotland-51967295; Harte, A. "A man tried to choke me during sex without warning." BBC News, November 28, 2019. https://www.bbc.com/news/uk-50546184.

48. There are many texts to support this claim. Two are: Littman, L., 2018. Rapid-onset gender dysphoria in adolescents and young adults: A study of parental reports. *PloS One,* 13(8): e0202330; Shrier, A., 2020. *Irreversible Damage: The Transgender Craze Seducing our Daughters.* Washington, D.C.: Regnery Publishing.

49. See, for example, Hayes, T. B., et al., 2002. Hermaphroditic, demasculinized frogs after exposure to the herbicide atrazine at low ecologically relevant doses. *Proceedings of the National Academy of Sciences,* 99(8): 5476–5480; Reeder, A. L., et al., 1998. Forms and prevalence of intersexuality and effects of environmental contaminants on sexuality in cricket frogs (*Acris crepitans*). *Environmental Health Perspectives,* 106(5): 261–266.

Chapter 8: Parenthood and Relationship

1. As contrasted with precociality, which is early self-sufficiency of hatchling and newborns, and linguistically related to, but not the same as, our concept of the "precocious" child.

2. Cornwallis, C. K., et al., 2010. Promiscuity and the evolutionary transition to complex societies. *Nature,* 466(7309): 969–972.

3. For the classic paper on how the distribution of resources in space and time affects mating systems, see Emlen, S. T., and Oring, L. W., 1977. Ecology, sexual selection, and the evolution of mating systems. *Science*, 197(4300): 215–223.

4. Madge, S., and Burn, H. 1988. *Waterfowl: An Identification Guide to the Ducks, Geese, and Swans of the World*. Boston: Houghton Mifflin.

5. Larsen, C. S., 2003. Equality for the sexes in human evolution? Early hominid sexual dimorphism and implications for mating systems and social behavior. *Proceedings of the National Academy of Sciences*, 100(16): 9103–9104.

6. Schillaci, M. A., 2006. Sexual selection and the evolution of brain size in primates. *PLoS One*, 1(1): e62.

7. von Bayern, A. M., et al., 2007. The role of food- and object-sharing in the development of social bonds in juvenile jackdaws (*Corvus monedula*). *Behaviour*, 144(6): 711–733.

8. Holmes, R. T., 1973. Social behaviour of breeding western sandpipers *Calidris mauri*. *Ibis*, 115(1): 107–123.

9. Rogers, W., 1988. Parental investment and division of labor in the Midas cichlid (*Cichlasoma citrinellum*). *Ethology*, 79(2): 126–142.

10. Eisenberg, J. F., and Redford, K. H., 1989. *Mammals of the Neotropics, Volume 2: The Southern Cone: Chile, Argentina, Uruguay, Paraguay*. Chicago: University of Chicago Press.

11. Haig, D., 1993. Genetic conflicts in human pregnancy. *Quarterly Review of Biology*, 68(4): 495–532.

12. Emlen and Oring, Ecology, sexual selection, and the evolution of mating systems.

13. Tertilt, M., 2005. Polygyny, fertility, and savings. *Journal of Political Economy*, 113(6): 1341–1371.

14. Insel, T. R., et al., 1998. "Oxytocin, Vasopressin, and the Neuroendocrine Basis of Pair Bond Formation." In *Vasopressin and Oxytocin*, Zingg, H. H., et al., eds. New York: Plenum Press, 215–224.

15. Ricklefs, R. E., and Finch, C. E., 1995. *Aging: A Natural History*. New York: Scientific American Library.

16. Personal communication from George Estabrook, 1997. Also see his paper: Estabrook, G. F., 1998. Maintenance of fertility of shale soils in a traditional agricultural system in central interior Portugal. *Journal of Ethnobiology*, 18(1): 15–33.

17. Maiani, G. *Tsunami: Interview with a Moken of Andaman Sea*. January 2006. http://www.maiani.eu/video/moken/moken.asp?lingua=en.

18. Some of the growing evidence of early domestication of dogs is in these two papers: Freedman, A. H., et al., 2014. Genome sequencing highlights the dynamic early history of dogs. *PLoS Genetics*, 10(1): e1004016; Bergström, A., et al., 2020. Origins and genetic legacy of prehistoric dogs. *Science*, 370(6516): 557–564.

19. de Waal, F., 2019. *Mama's Last Hug: Animal Emotions and What They Tell Us about Ourselves*. New York: W. W. Norton.

20. Palmer, B., 1998. The influence of breastfeeding on the development of the oral cavity: A commentary. *Journal of Human Lactation*, 14(2): 93–98.
21. Credit to our student Josie Jarvis for developing this hypothesis.

Chapter 9: Childhood

1. de Waal, F., 2019. *Mama's Last Hug: Animal Emotions and What They Tell Us about Ourselves*. New York: W. W. Norton, 97.
2. Fraser, O. N., and Bugnyar, T., 2011. Ravens reconcile after aggressive conflicts with valuable partners. *PLoS One*, 6(3): e18118.
3. Kawai, M., 1965. Newly-acquired pre-cultural behavior of the natural troop of Japanese monkeys on Koshima Islet. *Primates*, 6(1): 1–30.
4. "The blankest slates" is a phrase that first emerged in a program that Bret was teaching, from one of his students.
5. Both Asian and African elephants have an age of first reproduction that is similar to that of humans, but their age of independence—by some measures the end of childhood—is far earlier: five and eight years, respectively. No other animals—such as great apes, dolphins, parrots—come close.
6. It is also true that it is currently fashionable to raise children to be multilingual, but we might ask what the costs are. The social benefits are clear, but forcing the brain to maintain more linguistic competency and complexity than it would have historically must come with a trade-off.
7. Benoit-Bird, K. J., and Au, W. W., 2009. Cooperative prey herding by the pelagic dolphin, *Stenella longirostris*. *Journal of the Acoustical Society of America*, 125(1): 125–137.
8. Rutz, C., et al., 2012. Automated mapping of social networks in wild birds. *Current Biology*, 22(17): R669–R671.
9. Goldenberg, S. Z., and Wittemyer, G., 2020. Elephant behavior toward the dead: A review and insights from field observations. *Primates*, 61(1): 119–128.
10. Sutherland, W. J., 1998. Evidence for flexibility and constraint in migration systems. *Journal of Avian Biology*, 29(4): 441–446.
11. In this regard, childhood is rather like sexual reproduction. Both are adaptive responses to a changing world.
12. Lancy, D. F., 2014. *The Anthropology of Childhood: Cherubs, Chattel, Changelings*. Cambridge: Cambridge University Press, 209–212.
13. Gray, P., and Feldman, J., 2004. Playing in the zone of proximal development: Qualities of self-directed age mixing between adolescents and young children at a democratic school. *American Journal of Education*, 110(2): 108–146. Also Peter Gray, personal communication, September 2020.
14. See for instance the account of young children in the South Pacific by researcher Mary Martini, as recounted in Gray, P., 2013. *Free to Learn: Why Unleashing the Instinct to Play Will Make Our Children Happier, More Self-reliant, and Better Students for Life*. New York: Basic Books, 208–209.

15. This manuscript went to press after more than a year of COVID-19-driven lock-downs, which meant no school, and no recess, for many children for a very long time. Compared to that, any form of play between children would be an improvement.

16. In contrast to most authoritative parenting books, this one is excellent: Skenazy, L., 2009. *Free-Range Kids: How to Raise Safe, Self-Reliant Children (Without Going Nuts with Worry)*. New York: John Wiley & Sons.

17. The range of possible phenotypes that can be produced by a single genotype is called the reaction norm.

18. West-Eberhard, M. J., 2003. *Developmental Plasticity and Evolution*. New York: Oxford University Press, 41.

19. Lieberman, D., 2014. *The Story of the Human Body: Evolution, Health, and Disease*. New York: Vintage, 163.

20. See, for example, Pfennig, D. W., 1992. Polyphenism in spadefoot toad tadpoles as a locally adjusted Evolutionarily Stable Strategy. *Evolution*, 46(5): 1408–1420, and indeed everything out of the Pfennig lab: https://www.davidpfenniglab.com /spadefoots.

21. Mariette, M. M., and Buchanan, K. L., 2016. Prenatal acoustic communication programs offspring for high posthatching temperatures in a songbird. *Science*, 353(6301): 812–814.

22. West-Eberhard, *Developmental Plasticity and Evolution* 50–55.

23. Plasticity can take many forms. One is the uncoupling of morphological development from reproductive development—many salamanders retain larval characteristics as reproductive adults, keeping gills and webbed feet if the ecological conditions are better for them in water than on land. Changes in timing reflect another kind of plasticity—the eggs of some tropical tree frogs hatch before their time, into tadpoles, if they receive a signal from their siblings that a snake is eating them. Embryonic crocodiles that experience either low or high temperatures in their eggs become females; at intermediate temperatures they become male. The sequential hermaphroditism of many reef fish, such that many individuals have been both adult female and male before they die, is another form of plasticity. Plants have tropisms—they grow toward light, against gravity, or in response to touch—and they flower when day length, temperature, or rainfall prompts them to. Plant tissues also tend to retain more plasticity than do those of animals, growing leaves into a light gap or roots into a patch of magnesium. Constraint forces the creation of opportunity.

24. Karasik, L. B., et al., 2018. The ties that bind: Cradling in Tajikistan. *PloS One*, 13(10): e0204428.

25. WHO Multicentre Growth Reference Study Group and de Onis, M., 2006. WHO Motor Development Study: Windows of achievement for six gross motor development milestones. *Acta paediatrica*, 95, supplement 450: 86–95.

26. For an excellent popular account, see Gupta, S., September 14, 2019. Culture helps shape when babies learn to walk. *Science News*, 196(5).

27. Kenyan mothers actively teach their babies to sit, and then to walk: Super, C. M., 1976. Environmental effects on motor development: The case of "African infant precocity." *Developmental Medicine & Child Neurology*, 18(5): 561–567.

28. Taleb, N. N., 2012. *Antifragile: How to Live in a World We Don't Understand*, vol. 3. London: Allen Lane.

29. Wilcox, A. J., et al., 1988. Incidence of early loss of pregnancy. *New England Journal of Medicine*, 319(4): 189–194; Rice, W. R., 2018. The high abortion cost of human reproduction. *bioRxiv* (preprint). https://doi.org/10.1101/372193.

30. This is a fascinating account of the history of attachment theory: Bretherton, I., 1992. The origins of attachment theory: John Bowlby and Mary Ainsworth. *Developmental Psychology*, 28(5): 759–775.

31. As mentioned in the previous chapter, with regard to genomic imprinting. See Haig, D., 1993. Genetic conflicts in human pregnancy, *Quarterly Review of Biology*, 68(4): 495–532.

32. Trivers, R. L., 1974. Parent-offspring conflict. *Integrative and Comparative Biology*, 14(1): 249–264.

33. Spinka, M., Newberry, R. C., and Bekoff, M., 2001. Mammalian play: Training for the unexpected. *Quarterly Review of Biology*, 76(2): 141–168.

34. De Oliveira, C. R., et al., 2003. Play behavior in juvenile golden lion tamarins (Callitrichidae: Primates): Organization in relation to costs. *Ethology*, 109(7): 593–612.

35. Gray, P., 2011. The special value of children's age-mixed play. *American Journal of Play*, 3(4): 500–522.

36. See the CDC's Autism and Developmental Disabilities Monitoring (ADDM) Network site: https://www.cdc.gov/ncbddd/autism/addm.html.

37. Cheney, D. L., and Seyfarth, R. M., 2007. *Baboon Metaphysics: The Evolution of a Social Mind*. Chicago: University of Chicago Press, 155, 176–177, 197.

38. Whitaker, R., 2015. *Anatomy of an Epidemic: Magic Bullets, Psychiatric Drugs, and the Astonishing Rise of Mental Illness in America*. 2nd ed. New York: Broadway Books. See, in particular, chapter 11: "The Epidemic Spreads to Children."

39. See, for instance, this fantastic analysis: Sommers, C. H., 2001. *The War against Boys: How Misguided Feminism Is Harming Our Young Men*. New York: Simon & Schuster.

40. For example, left-handers win more fights than do right-handers: Richardson, T., and Gilman, T., 2019. Left-handedness is associated with greater fighting success in humans. *Scientific Reports*, 9(1): 1–6.

41. Developmental psychologist Jean Piaget was the first to demonstrate that children grasp rules better when playing on their own than when actively directed by adults. Piaget, J., 1932. *The Moral Judgment of the Child*. Reprint ed. 2013. Abingdon-on-Thames, UK: Routledge.

42. Frank, M. G., Issa, N. P., and Stryker, M. P., 2001. Sleep enhances plasticity in the developing visual cortex. *Neuron*, 30(1): 275–287.

Chapter 10: School

1. Lancy, D. F., 2015. *The Anthropology of Childhood: Cherubs, Chattel, Changelings,* 2nd ed. Cambridge: Cambridge University Press, 327–328.
2. Gatto, J. T., 2001. *A Different Kind of Teacher: Solving the Crisis of American Schooling.* Berkeley: Berkeley Hills Books.
3. From map on page 4 of Finer, M., et al., 2009. Ecuador's Yasuni Biosphere Reserve: A brief modern history and conservation challenges. *Environmental Research Letters,* 4(3): 034005.
4. Heying, H., 2019. "The Boat Accident." Self-published on Medium. https://medium .com/@heyingh.
5. Definition of teaching: When individual A modifies their behavior only in the presence of naive individual B, at a cost or no immediate benefit to A, such that B acquires knowledge earlier or more efficiently or faster than it might otherwise. From Caro, T. M., and Hauser, M. D., 1992. Is there teaching in nonhuman animals? *Quarterly Review of Biology,* 67(2): 151–174.
6. Leadbeater, E., and Chittka, L., 2007. Social learning in insects—from miniature brains to consensus building. *Current Biology,* 17(16): R703–R713.
7. Franks, N. R., and Richardson, T., 2006. Teaching in tandem-running ants. *Nature,* 439(7073): 153.
8. Thornton, A., and McAuliffe, K., 2006. Teaching in wild meerkats. *Science,* 313(5784): 227–229.
9. Bender, C. E., Herzing, D. L., and Bjorklund, D. F., 2009. Evidence of teaching in Atlantic spotted dolphins (*Stenella frontalis*) by mother dolphins foraging in the presence of their calves. *Animal Cognition,* 12(1): 43–53.
10. Many of these examples (e.g., from cats and primates) are reviewed in Hoppitt, W. J., et al., 2008. Lessons from animal teaching. *Trends in Ecology & Evolution,* 23(9): 486–493.
11. Hill, J. F., and Plath, D. W., 1998. "Moneyed Knowledge: How Women Become Commercial Shellfish Divers." In *Learning in Likely Places: Varieties of Apprenticeship in Japan,* Singleton, J., ed. Cambridge: Cambridge University Press, 211–225.
12. Lancy, *Anthropology of Childhood,* 209–212.
13. See, for instance, Lake, E., 2014. Beyond true and false: Buddhist philosophy is full of contradictions. Now modern logic is learning why that might be a good thing. *Aeon,* May 5, 2014. https://aeon.co/essays/the-logic-of-buddhist-philosophy-goes -beyond-simple-truth.
14. Borges, J. L., 1944. *Funes the Memorious.* Reprinted in several collections, including Borges, J. L., 1964. *Labyrinths: Selected Stories and Other Writings.* New York: New Directions.
15. Gatto, J. T., 2010. *Weapons of Mass Instruction: A Schoolteacher's Journey through the Dark World of Compulsory Schooling.* Gabriola Island: New Society Publishers.
16. As posed by Derrick Jensen in his 2004 book *Walking on Water: Reading, Writing, and Revolution.* White River Junction, VT: Chelsea Green Publishing, 41.

17. A brief précis of the adaptive landscapes metaphor can be found in endnote 19 of chapter 3.

18. For the classic analysis of paradigm shifts, see Kuhn, T. S., 1962. *The Structure of Scientific Revolutions*. Chicago: University of Chicago Press.

19. Müller, J. Z., 2018. *The Tyranny of Metrics*. Princeton, NJ: Princeton University Press. See especially chapter 7, "Colleges and Universities," 67–88, and chapter 8, "School," 89–102.

20. See our co-written essay: Heying, H. E., and Weinstein, B., 2015. "Don't Look It Up," *Proceedings of the 2015 Symposium on Field Studies at Colorado College*, 47–49. https://www.academia.edu/35652813/Dont_Look_It_Up.

21. Quotation from profile of Teller in Lahey, J., 2016. Teaching: Just like performing magic. *Atlantic*, January 21, 2016. https://www.theatlantic.com/education/archive/2016/01/what-classrooms-can-learn-from-magic/425100.

22. The adaptive landscape metaphor applies to learning as well: Once at some adaptive peak, it is nearly impossible, in analytic or social space, to move down off that peak—to a less adapted form—even if you can see a higher peak nearby. Those who are entering the landscape anew will rise to some peak that they are near, without the constraints of already being at some local peak. Those who are already on the map are already stable.

23. Heying, H., 2019. On college presidents. *Academic Questions*, 32(1): 19–28.

24. Haidt, J. "How two incompatible sacred values are driving conflict and confusion in American universities." Lecture, Duke University, Durham, NC, October 6, 2016.

25. Heying, H. "Orthodoxy and heterodoxy: A conflict at the core of education." Invited talk, Academic Freedom Under Threat: What's to Be Done?, Pembroke College, Oxford University, May 9–10, 2019.

Chapter 11: Becoming Adults

1. As recounted in McWhorter, L. V., 2008. *Yellow Wolf, His Own Story*. Caldwell, ID: Caxton Press, 297–300. Originally published in 1940.

2. Markstrom, C. A., and Iborra, A., 2003. Adolescent identity formation and rites of passage: The Navajo Kinaalda ceremony for girls. *Journal of Research on Adolescence*, 13(4): 399–425.

3. Becker, A. E., 2004. Television, disordered eating, and young women in Fiji: Negotiating body image and identity during rapid social change. *Culture, Medicine and Psychiatry*, 28(4): 533–559.

4. For two excellent descriptions of how post-modernism, and its intellectual descendants such as post-structuralism and Critical Race Theory, have invaded the academy, see Pluckrose, H., Lindsay, J. and Boghossian, P., 2018. Academic grievance studies and the corruption of scholarship. *Areo*, February 10, 2018; and Pluckrose, H., and Lindsay, J., 2020. *Cynical Theories: How Activist Scholarship Made Everything about Race, Gender, and Identity—and Why This Harms Everybody*. Durham, NC: Pitchstone Publishing.

5. There are many accounts of how postmodern-inspired activism has torn asunder good systems. Here are just a few: Murray, D., 2019. *The Madness of Crowds: Gender, Race and Identity.* London: Bloomsbury Publishing; Daum, M., 2019. *The Problem with Everything: My Journey through the New Culture Wars.* New York: Gallery Books; Asher, L., 2018. How Ed schools became a menace. *The Chronicle of Higher Education,* April 2018.

6. Dawkins, R., 1998. Postmodernism disrobed. *Nature,* 394(6689): 141–143.

7. Inroads are being made into sport, however, via bullying and expectations of social conformity, in the form of Trans Rights Activists (not to be confused with actual trans people), who have effected changes in several sports to allow natal men to compete in women's sport, which is patently unfair and unsportsmanlike. See Hilton, E. N., and Lundberg, T. R., 2021. Transgender women in the female category of sport: Perspectives on testosterone suppression and performance advantage. *Sports Medicine,* 51(2021): 199–214.

8. Crawford, M. B., 2015. *The World Beyond Your Head: On Becoming an Individual in an Age of Distraction.* New York: Farrar, Straus and Giroux, 48–49.

9. Heying, H., 2018. "Nature Is Risky. That's Why Students Need It." *New York Times,* April 30, 2018. https://www.nytimes.com/2018/04/30/opinion/nature-students-risk .html.

10. Lukianoff, G., and Haidt, J., 2019. *The Coddling of the American Mind: How Good Intentions and Bad Ideas Are Setting Up a Generation for Failure.* New York: Penguin Books.

11. Estabrook, G. F., 1994. Choice of fuel for bagaco stills helps maintain biological diversity in a traditional Portuguese agricultural system. *Journal of Ethnobiology,* 14(1): 43–57.

12. Again: Heying, H., 2019. "The Boat Accident." Self-published on Medium. https:// medium.com/@heyingh.

13. For a somewhat more complete take on this, we recommend our December 12, 2017, article in the *Washington Examiner* ("Bonfire of the Academies: Two Professors on How Leftist Intolerance Is Killing Higher Education"); and on YouTube, both Mike Nayna's three-part documentary and Benjamin Boyce's exhaustive multipart series on the breakdown at Evergreen.

14. As first described by Richard D. Alexander in his book *The Biology of Moral Systems.* Hawthorne, NY: Aldine de Gruyter, 1987.

15. Lahti, D. C., and Weinstein, B. S., 2005. The better angels of our nature: Group stability and the evolution of moral tension. *Evolution and Human Behavior,* 26(1): 47–63.

16. Cheney, D. L., and Seyfarth, R. M., 2007. *Baboon Metaphysics: The Evolution of a Social Mind.* Chicago: University of Chicago Press.

17. Brosnan, S. F., and de Waal, F. B., 2003. Monkeys reject unequal pay. *Nature,* 425(6955): 297–299.

18. Adams, J., et al., 1999. National household survey on drug abuse data collection. Final report, as cited in Green, T., Gehrke, B., and Bardo, M., 2002. Environmental

enrichment decreases intravenous amphetamine self-administration in rats: Dose-response functions for fixed- and progressive-ratio schedules. *Psychopharmacology*, 162(4): 373–378.

19. Bardo, M., et al., 2001. Environmental enrichment decreases intravenous self-administration of amphetamine in female and male rats. *Psychopharmacology*, 155(3): 278–284.

20. Tristan Harris has been sounding this alarm for years. Here is an account from 2016: Bosker, B., 2016. The binge breaker: Tristan Harris believes Silicon Valley is addicting us to our phones: He's determined to make it stop. *Atlantic*, November 2016. https://www.theatlantic.com/magazine/archive/2016/11/the-binge-breaker/501122. Also listen to Tristan's conversation with Bret on *DarkHorse* podcast, aired February 25, 2021.

Chapter 12: Culture and Consciousness

1. In his 1974 article, What is it like to be a bat? *Philosophical Review*, 83(4): 435–450, Thomas Nagel suggests that the sign of a conscious mind is one that can consider itself. Our formulation extends this but does not contradict this formulation. We add: the conscious mind, having considered itself, can communicate that consideration to others of its own kind.

2. Cheney, D. L., and Seyfarth, R. M., 2007. *Baboon Metaphysics: The Evolution of a Social Mind*. Chicago: University of Chicago Press.

3. In fact, there is evidence of at least two, perhaps more, independent origins of farming in China alone—rice in the humid south, and millet in the colder, arid north. See Barton, L., et al., 2009. Agricultural origins and the isotopic identity of domestication in northern China. *Proceedings of the National Academy of Sciences*, 106(14): 5523–5528.

4. This is an accessible overview of Asch's original conformity experiments, and related work: Asch, S. E., 1955. Opinions and social pressure. *Scientific American*, 193(5): 31–35.

5. Mori, K., and Arai, M., 2010. No need to fake it: Reproduction of the Asch experiment without confederates. *International Journal of Psychology*, 45(5): 390–397.

6. Morales, H., and Perfecto, I., 2000. Traditional knowledge and pest management in the Guatemalan highlands. *Agriculture and Human Values*, 17(1): 49–63.

7. Estabrook, G. F., 1994. Choice of fuel for bagaco stills helps maintain biological diversity in a traditional Portuguese agricultural system. *Journal of Ethnobiology*, 14(1): 43–57.

8. Boland, M. R., et al., 2015. Birth month affects lifetime disease risk: A phenome-wide method. *Journal of the American Medical Informatics Association*, 22(5): 1042–1053. There is also abundant other research looking at birth month effects on health and physiology, including one finding a clear link between birth month and myopia: Mandel, Y., et al., 2008. Season of birth, natural light, and myopia. *Ophthalmology*, 115(4): 686–692.

9. Smith, N. J. H., 1981. *Man, Fishes, and the Amazon*. New York: Columbia University Press, 87.

10. Ruud, J., 1960. *Taboo: A Study of Malagasy Customs and Beliefs*. Oslo: Oslo University Press, 109. Ruud calls it a "tufted umbrette," but this species is more usually referred to as a hamerkop.

11. Ruud, *Taboo*. Mutton, 85; hedgehogs, 239; pumpkin, 242; house construction, 120.

12. As cited in a roundabout way in Ruud, *Taboo*, 1.

13. Ruud, *Taboo*. Landslide, 115; rabies, 87; divorce, 246.

14. Campbell, J. *The Hero's Journey: Joseph Campbell on His Life and Work*. Novato, CA: New World Library, 90.

15. Ehrenreich, B., 2007. *Dancing in the Streets: A History of Collective Joy*. New York: Metropolitan Books.

16. Chen, Y., and VanderWeele, T. J., 2018. Associations of religious upbringing with subsequent health and well-being from adolescence to young adulthood: An outcome-wide analysis. *American Journal of Epidemiology*, 187(11): 2355–2364.

17. Whitehouse, H., et al., 2019. Complex societies precede moralizing gods throughout world history. *Nature*, 568(7751): 226–299.

18. Hammerschlag, C. A., 2009. The Huichol offering: A shamanic healing journey. *Journal of Religion and Health*, 48(2): 246–258.

19. Bye, R. A., Jr., 1979. Hallucinogenic plants of the Tarahumara. *Journal of Ethnopharmacology*, 1(1979): 23–48.

Chapter 13: The Fourth Frontier

1. Mann, C. C., 2005. *1491: New Revelations of the Americas before Columbus*. New York: Alfred A. Knopf.

2. Cabodevilla, M. Á., 1994. Los Huaorani en la historia de los pueblos del Oriente. Cicame; as cited by Finer, M., et al., 2009. Ecuador's Yasuní Biosphere Reserve: A brief modern history and conservation challenges. *Environmental Research Letters*, 4(2009): 1–15.

3. Williams, G. C., 1957. Pleiotropy, natural selection, and the evolution of senescence. *Evolution*, 11(4): 398–411; Weinstein, B. S., and Ciszek, D., 2002. The reserve-capacity hypothesis: Evolutionary origins and modern implications of the trade-off between tumor-suppression and tissue-repair. *Experimental Gerontology*, 37(5): 615–627.

4. Dunning, N. P., Beach, T. P., and Luzzadder-Beach, S., 2012. Kax and kol: Collapse and resilience in lowland Maya civilization. *Proceedings of the National Academy of Sciences*, 109(10): 3652–3657.

5. Beach, T., et al., 2006. Impacts of the ancient Maya on soils and soil erosion in the central Maya Lowlands. *Catena*, 65(2): 166–178.

6. Wright, R., 2001. *Nonzero: The Logic of Human Destiny*. New York: Vintage.

7. Blake, J. G., and Loiselle, B. A., 2016. Long-term changes in composition of bird communities at an "undisturbed" site in eastern Ecuador. *Wilson Journal of Ornithology*, 128(2): 255–267.

8. Boyd, J. P., et al., 1950. *The Papers of Thomas Jefferson*, 33 vols. Princeton, NJ: Princeton University Press.

9. Alexander, R. D., 1990. *How Did Humans Evolve? Reflections on the Uniquely Unique Species*. Ann Arbor, MI: Museum of Zoology, University of Michigan. Special Publication No. 1.

Glossary

1. Wright, S. 1932. The roles of mutation, inbreeding, crossbreeding and selection in evolution. *Proceedings of the Sixth International Congress of Genetics*, 1: 356–366.

2. Taleb, N. N., 2012. *Antifragile: How to Live in a World We Don't Understand* (vol. 3). London: Allen Lane.

3. Chesterton, G. K., 1929. "The Drift from Domesticity." In *The Thing*. Aeterna Press.

Index

Page numbers in italics refer to illustrations.